M000087330

ANIMALADIES

ANIMALADIES

Gender, Animals, and Madness

Edited by

Lori Gruen and Fiona Probyn-Rapsey

BLOOMSBURY ACADEMIC

NEW YORK • LONDON • OXFORD • NEW DELHI • SYDNEY

BLOOMSBURY ACADEMIC
Bloomsbury Publishing Inc
1385 Broadway, New York, NY 10018, USA
50 Bedford Square, London, WC1B 3DP, UK

BLOOMSBURY, BLOOMSBURY ACADEMIC and the Diana logo are
trademarks of Bloomsbury Publishing Plc

First published in the United States of America 2019

Cover design: Eleanor Rose
Cover image: lynn mowson, *boobscape 2016–18*, latex,
tissue and string, (detail). Documentation: Kerry Leonard

Library of Congress Cataloging-in-Publication Data
Names: Gruen, Lori, editor. | Probyn-Rapsey, Fiona, editor.
Title: Animaladies : gender, animals and madness / edited by
Lori Gruen and Fiona Probyn-Rapsey.
Description: New York, NY : Bloomsbury, [2018] |
Includes bibliographical references and index.
Identifiers: LCCN 2018018008 (print) | LCCN 2018029477 (ebook) |
ISBN 9781501342172 (ePDF) | ISBN 9781501342165 (ePUB) |
ISBN 9781501342158 (hardback : alk. paper)
Subjects: LCSH: Human-animal relationships | Women and animals. |
Animals–Social aspects. | Zoophilia. Classification: LCC QL85 (ebook) |
LCC QL85 .A5213 2018 (print) | DDC 591.5–dc23
LC record available at https://lccn.loc.gov/2018018008

ISBN: HB: 978-1-5013-4215-8
         ePDF: 978-1-5013-4217-2
         eBook: 978-1-5013-4216-5

Typeset by Integra Software Services Pvt. Ltd.
Printed and bound in the United States of America

To find out more about our authors and books visit www.bloomsbury.com
and sign up for our newsletters.

CONTENTS

FIGURES

NOTES ON CONTRIBUTORS

Nekeisha Alayna Alexis is an independent scholar with wide-ranging interests connected to human and other animal oppression and liberation. She received her Bachelor of Arts degree from New York University, New York, and her Masters of Arts: Theological Studies degree from Anabaptist Mennonite Biblical Seminary (AMBS), Elkhart, Indiana. She is currently the Intercultural Competence and Undoing Racism coordinator at AMBS, where she assists the institution in its strategic goals in this area, as well as works on the Communication team as a graphic designer and website specialist. Nekeisha has presented and published extensively in both theological and religious settings on a number of topics such as animal liberation, veganism, and Christian peacemaking; the interconnected nature of human violence toward other animals and violence among human selves; intersecting oppressions of race, gender, and/or species; and anarchist politics and alternative Christian faith and ethics. Her most recent writings include "But Isn't All of Creation Violent?" in *A Faith Encompassing All of Creation: Addressing Commonly Asked Questions about Christian Care of the Environment* (Cascade Books, 2014) and "Table Talk: The Violence of Grace and Gratitude" (*The Mennonite*, September 2016). She is motivated by the hope and trust that her intellectual efforts around compassion will inspire others to transform their relationships to other creatures. She dedicates her work toward that end to her loved ones, Mocha and Cairo.

Liz Bowen is a Ph.D. candidate in English and comparative literature at Columbia University, where she works at the intersections of 20th and 21st century American literature, disability studies, and critical animal studies. Her dissertation, "Animal Abilities: Disability, Species Difference, and Aesthetic Innovation in the Long 20th Century," traces the intertwined deployments of disability, animality, and cognitive otherness as sites for literary experimentation, from Faulkner to the Harlem Renaissance to contemporary poetics. Liz has been the graduate organizer for Columbia's University Seminar on Disability, Culture, and Society for the past three years, and recently taught her first course on literary animal studies. She is also a widely published poet and critic, whose work has appeared or is forthcoming in *Boston Review, Lit Hub, The New Inquiry,* the *Journal of Literary and Cultural Disability Studies,* and *Humanimalia: A Journal of Human/Animal Interface Studies.*

Alice Crary is Professor of Philosophy in the Graduate Faculty and Co-Chair of the Gender and Sexuality Studies program of The New School for Social

Research (NSSR) in New York City. She is well known for her numerous scholarly works on the moral dimension of language, as well as edited collections on Wittgenstein, Cora Diamond, and Stanley Cavell. Crary is the author of two monographs on ethics, *Beyond Moral Judgment* (Harvard, 2007) and *Inside Ethics: On the Demands of Moral Thought* (Harvard, 2016).

Heather Fraser is Associate Professor in Social Science in the School of Public Health and Social Work, Health Faculty, QUT, Brisbane (http://staff.qut.edu.au/staff/fraserh2/). Heather teaches units relating to critical social work, helping alliances and 'addictions'. With (A/Prof.) Nik Taylor she authored the (2016) book *Neoliberalization, Universities and the Public Intellectual: Species, Gender and Class and the Production of Knowledge* (Palgrave, London).

Kathryn Gillespie earned her Ph.D. in Geography from the University of Washington (UW) in 2014. From 2016 to 2018 she was a postdoctoral fellow in Animal Studies at Wesleyan University. Her current research sits at the nexus of critical race theory (especially black feminisms), postcolonial studies, and critical animal studies, focusing on gendered human–animal relations and racialized histories of coloniality at the Louisiana State Penitentiary farm and rodeo at Angola. Her dissertation research examined the gendered commodification of the lives and bodies of animals in the dairy industry in the Pacific Northwestern United States. This work became *The Cow with Ear Tag #1389*, University of Chicago Press, 2018. She has published in scholarly journals, such as *Hypatia* and *Gender, Place, and Culture*, and has coedited two books, *Critical Animal Geographies: Politics, Intersections and Hierarchies in a Multispecies World* (Routledge 2015, with Rosemary-Claire Collard) and *Economies of Death: Economic Logics of Killable Life and Grievable Death* (Routledge 2015, with Patricia J. Lopez).

Lori Gruen is the William Griffin Professor of Philosophy at Wesleyan University. She is also Professor of Feminist, Gender, and Sexuality Studies and Coordinator of Wesleyan Animal Studies. She is the author and editor of ten books, including *Ethics and Animals: An Introduction* (Cambridge, 2011), *Reflecting on Nature: Readings in Environmental Philosophy and Ethics* (Oxford, 2012), *Ethics of Captivity* (Oxford, 2014), *Entangled Empathy* (Lantern, 2015), and *Critical Terms for Animal Studies* (University of Chicago, 2018). Her work in practical ethics focuses on issues that impact those often overlooked in traditional ethical investigations, for example women, people of color, incarcerated people, and nonhuman animals. She is a former editor of *Hypatia: A Journal of Feminist Philosophy*, is Fellow of the Hastings Center for Bioethics, is Faculty Fellow at Tufts' Cummings School of Veterinary Medicine's Center for Animals and Public Policy, was the Spring 2018 Laurance S. Rockefeller Visiting Professor for Distinguished Teaching at Princeton University, and was the first chair of the Faculty Advisory Committee of the Center for Prison Education at Wesleyan.

pattrice jones cofounded VINE Sanctuary, an LGBTQ-run farmed animal sanctuary that operates within an understanding of the intersection of oppressions. She is the author of *Aftershock: Confronting Trauma in a Violent World: A Guide for Activists and Their Allies* (Lantern, 2007) and *The Oxen at the Intersection* (Lantern, 2014), and many chapters and essays.

Hannah Monroe holds a Master of Arts in Critical Sociology from Brock University, where her thesis research focused on the experiences of autistic people, employing post-structural analysis. She currently supports and mentors autistic children. She plans to complete a Master of Social Work in order to become a therapist, with the goal of empowering neurodiverse young people. Her other publications include a chapter titled "Post-Structural Analyses of Conformity and Oppression: A Discussion of Critical Animal Studies and Neurodiversity" in *Animals, Disability and the End of Capitalism: Voices from the Eco-ability Movement*, edited by Anthony J. Nocella, John Lupinacci, and Amber E. George. She has spoken on the intersections between neurodiversity and critical animal studies at the Decolonizing Critical Animal Studies, Cripping Critical Animal Studies conference in June 2016, and at the third Annual Eco-Ability Conference in December 2015.

lynn mowson is a practicing sculptor and animal advocate. Her practice-led Ph.D., entitled "*beautiful little dead things: empathy, witnessing, trauma and animals' suffering,*" was completed at VCA, The University of Melbourne, in 2015. Her sculptural research is featured in the book *The Art of the Animal* (Lantern Press, 2015) and the exhibition 'SPOM: The Sexual Politics of Meat' at The Animal Museum, Los Angeles, held in February 2017. lynn is Research Assistant for the Human Rights and Animal Ethics Research Network (HRAE), University of Melbourne, is on the committee for HRAE, is Vice-Chair of the Australasian Animal Studies Association, and is Associate for the New Zealand Centre for Human Animal Studies. Website: https://lynnmowson.com/

Fiona Probyn-Rapsey is Professor in the School of Humanities and Social Inquiry at the University of Wollongong, Australia. She is the author and coeditor of three books, including *Made to Matter* (Sydney University Press, 2013), *Animal Death* (Sydney University Press, 2013), and *Animals in the Anthropocene: Critical Perspectives on Non-human Futures* (SUP, 2015). Fiona is also Series Editor (with Melissa Boyde) of the Animal Publics book series through Sydney University Press (http://sydney.edu.au/sup/about/animal_publics.html).

Guy Scotton is an independent researcher, currently exploring the role of emotions and virtues in theories of interspecies justice. He is an editor of the open access journal *Politics and Animals* (www.politicsandanimals.org/).

Hayley Singer received her Ph.D. in Creative Writing from the University of Melbourne, where she teaches in the School of Culture and Communication. Her creative and scholarly writing practices engage with eco/feminism's intersectional politics and poetics. Her fiction, nonfiction, poetry, and book reviews have been published in Australian journals and anthologies, including *Meanjin, Page Seventeen, Writing from Below, Press: 100 Love Letters,* and *Animal Studies Journal.*

James Stanescu is Assistant Professor of Communication Studies at Mercer University, in Macon, Georgia. He is the editor, with Kevin Cummings, of *The Ethics and Rhetoric of Invasion Ecology* (Lexington, 2016). Stanescu publishes frequently on issues surrounding animal and environmental philosophy, and is finishing a monograph on the genealogy of the factory farm. You can also find his occasional writings at his blog, *Critical Animal.*

Vasile Stanescu received his Ph.D. in the interdisciplinary program of Modern Thought and Literature (MTL) at Stanford University. He currently serves as Assistant Professor and Director of the Program in Speech and Debate at Mercer University. Stanescu is the co-senior editor of the *Critical Animal Studies* book series. He is currently working on his forthcoming book project, *Happy Meals: Animals, Affect, and the Myth of Consent.*

Nik Taylor is Associate Professor of Sociology in the Department of Human Services and Social Work at the University of Canterbury, New Zealand. She is an award-winning author who has published four books, including *Humans, Animals and Society* (Lantern, 2012), and over fifty journal articles and book chapters on the human–pet bond; treatment of animals and animal welfare; links between human aggression and animal cruelty; slaughterhouses; meat-eating; and animal shelter work.

Yvette Watt is a Lecturer at the School of Creative Arts, University of Tasmania and Lead Researcher of the UTAS College of Arts and Law Animal Studies Theme Area (ASTA). Watt was a founding member of the Australasian Animal Studies Association and is a current committee member of Minding Animals International.

Watt's art practice spans 30 years and includes numerous solo and group exhibitions. She has been actively involved in animal advocacy since the mid 1980s, and her artwork is heavily informed by her activism. Her work is held in numerous public and private collections in Australia including Parliament House, Canberra, Artbank and the Art Gallery of Western Australia. Watt is a co-editor of and contributor to *Considering Animals: Contemporary Studies in Human-Animal Relations* (Ashgate, 2011). Other publications include "Down on the Farm: Why do Artists Avoid Farm Animals as Subject Matter?", in *Meat Cultures*, Annie Potts (ed), Brill (2016); 'Animal Factories: Exposing Sites of

Capture', in *Captured: Animals Within Culture*, Melissa Boyd, ed, (Palgrave McMillan, 2014) and 'Artists, Animals and Ethics', in *Antennae: the journal of nature in culture*, (issue 19, 2011).

Cheryl Wylie coordinates animal care at VINE sanctuary, an LGBTQ-led farmed animal refuge that was the first sanctuary to devise a method of rehabilitating roosters used in cockfighting. She is a veterinary technician who participated in 4-H and FFA as a child.

ACKNOWLEDGMENTS

We would like to begin by thanking all the courageous people who work to bring compassion to our mad world with all its animaladies.

We would like to thank HARN: Human Animal Research Network at the University of Sydney (especially Dinesh Wadiwel and Agata Mrva-Montoya, Charlie Jackson-Martin and Indi) for their help with the first Animaladies conference held there in 2016, where Lori Gruen was a keynote speaker. Colleagues from AASA: Australasian Animal Studies Association, Melissa Boyde, Madeleine Boyd, and Yvette Watt did an amazing job on the first Animaladies Exhibition held at Verge Gallery (Glebe), where we all got to see lynn mowson's incredible piece *boobscapes* in the flesh. Seeing it there "for real" inspired us to ask her if we could use an image of that piece for the front cover of this book. We're delighted that she agreed. Yvette Watt's work on *Duck Lake Project* also inspired so many of us during 2016 and 2017, and we are delighted to present part of her work in this volume. Brian Rapsey did a fantastic job (as always) filming and photographing the event and was also instrumental in our road trip to Val Plumwood's home at Plumwood Mountain, where we were warmly welcomed by Anne Edwards and Tash Fjin. Thanks are also due to Sunday and Olive and Brian who were squished in the back of the car and patiently indulged Lori and Fiona's never ending conversations on our road trip.

Fiona would also like to acknowledge the support and assistance of the School of Humanities and Social Inquiry at the University of Wollongong which is proving itself to be a great place to do Animal Studies work, especially with terrific colleagues like Melissa Boyde, Alison Moore, Denise Russell, Colin Salter, Esther Alloun, and Kathy Varvaro, who all just happen to be on the committee for the next Animaladies conference and members of the ASRN: Animal Studies Research Network. Lori would like to acknowledge the support of Wesleyan Animal Studies and especially thank pattrice jones for taking care of Taz and Zinnie while I was away in Australia.

We would like to thank all the contributors to this book, who all worked tirelessly incorporating suggestions from our reviewers (thanks to them also), as well as helping to make the idea of "animaladies" come to life in original and insightful ways. Sue Pyke did a brilliant job pulling the final manuscript together. Thanks to Carol J. Adams for her support for the project and for agreeing to write a postscript/discussion for us. We'd also like to thank Haaris Naqvi and Katherine De Chant at Bloomsbury, both of whom have been a pleasure to work with.

Working and writing and laughing together, both in person and across the world, has been a joy for us. Fiona would like to thank Lori for her generosity over the years, both as scholar and as a friend. I have come to appreciate her philosopher's eye, her feminist work ethic, and her ethic of care even more acutely in the process. Lori wants to express her deepest appreciation for Fiona's brilliance and good humor, only Fi could come up with "animaladies" and the energy and passion to initiate and sustain this critical conversation.

LG and FPR
March 1, 2018

DISTILLATIONS

Lori Gruen and Fiona Probyn-Rapsey

I felt a Funeral, in my Brain,
And Mourners to and fro
Kept treading—treading—till it seemed
That Sense was breaking through—

—Dickinson, 1896

"Animaladies" conjoins two words, "animal" and "maladies" or, looked at another way, "animal" and "ladies." Whichever two words you see when you look at it, it is always three: "animal," "malady," and "ladies," a triangulation within Animal Studies that draws gender, animals, and maladies together. Fiona originally coined the term to describe the dis-ease of current human–animal relationships, and the idea that acknowledging these maladies was a necessary catalyst for positive change (Probyn-Rapsey 2014, 4). In 2016, "Animaladies" became an interdisciplinary conference held at the University of Sydney, with Lori Gruen as the keynote speaker. The conference called for papers that interrogated the connections between gender, madness, and animality, bringing together a feminist archive on histories of madness and the flourishing field of Animal Studies.

We were blown away by the excitement that the call for papers for the conference and this volume generated. The term resonates with writers, scholars, activists, and artists whose various engagements with other animals and with Animal Studies are shaped in one way or another by animaladies. Defined broadly, animaladies are sites of tension produced by acknowledging how our relationships with other animals are damaged. These relationships are damaged in a variety of ways, both by common attitudes of human superiority and by the violent and disturbing implications of these attitudes. Naming these damaged relationships as animaladies helps us to see how we might reframe both our attitudes and their consequences within various social contexts. How we imagine the damage and what we do with that damage is a genuine question. The power of animaladies is that they usually resist attempts to disrupt them. Often those who try to escape the damage by reframing it are themselves identified as mad, *as* damaged, much like

Sara Ahmed's "feminist killjoy" (2017). In this volume the authors attempt to disentangle these damaged relations, in order to highlight possible escape routes.

Animaladies also express political and psychological discontent, familiar from feminist theorizing about violent systems of power. Animaladies highlight how pathologizing human–animal relationships blocks empathy toward animals because the characterization of animal advocacy as mad, "crazy," and feminized, distracts attention from broader social disorder regarding human exploitation of animal life. Understanding how the "madness" of our relationships with animals intersects with the "madness" of taking animals seriously is another major aim of this book. The volume will not attempt to show where madness truly lies, but rather illuminates how it is distributed; how it is made purposeful; how it is disguised; how it is made to work for social change or against it; and how it is shaped as an insult, embraced as a zone of quarantine or left as an undefined fear.

The concept of animaladies also builds on a long-standing feminist archive on the topic of women and madness. Elaine Showalter's *The Female Malady* (1985) deserves special mention. Showalter's early work situates madness as a specifically "female malady," enshrined by a patriarchal psychiatric tradition that creates a double bind for women—damned if they do conform to femininity and damned if they don't. As many feminists have argued (Chesler 1972; Ussher 1991; Russell 1995; Gilbert and Gubar 2000), if being a woman also approximates a maladjustment to patriarchal norms, then it is little surprise that women are more likely to experience "madness" and be seen as "mad." For Showalter, it is not only that more women are diagnosed with mental health problems (which they are still today), and that the psychiatric profession is male dominated (which it still is), but that the condition of madness is itself gendered, coded, as a *female* malady. Showalter identifies a Victorian fascination with three cultural icons of the "madwoman"—Ophelia, Crazy Jane, and Lucia—each of whom worked to establish "female sexuality and feminine nature as the source of the female malady" and each "stood for a different interpretation of woman's madness and man's relationship to it" (10). Linked to sexuality (excessive or frigid), reason (too much or not enough), obedience (too much or not enough), feminist work on madness emphasizes how cultural icons of the madwoman persist in different forms and preempt assumptions about (and the psychiatric and therapeutic treatment of) "mad" women. The hystericizing of women and the reclaiming of hysteria as a form of feminist resistance has never gone away. It helps us understand how animal advocates, whose objections extend to anthropocentrism, are also hystericized and pathologized as singular damaged individuals. This focus draws our attention away from normalized (and most often non-pathologized) racial and colonial violence that is also linked to gender and the exploitation of animals.

Pathologization is a mechanism for discriminating between acceptable and unacceptable behaviors, identities, and beliefs. In her historical analysis of the animal rights movement in the United States, Diane Beers shows that pathologization of animal advocacy was a key mechanism for disciplining the movement and, more specifically, the white women who constituted the majority of members. Beers writes that "several late nineteenth-century physicians concocted a diagnosable form of mental illness" called "zoophilpsychosis," an invention that she describes as the "most vicious of all" (2006, 109). According to the primary proponent of this diagnosis, physician Charles Dana, zoophilpsychosis was a condition produced by "fine feelings gone wrong" and "overgrown sentimentality" (1909, 382). He presents one case of a man who "worried so much" about horses and another case of a woman who was "victim of a cat obsession." The gender differences between the "cures" are telling—the man's feelings for horses have "gone wrong" and are fixed after three years of therapy while the woman is advised to have "gynecological treatment" to deal with her "perversion of instinct." Because the animal movement was (and still is) dominated by women, "doctors further concluded that the weaker sex was particularly susceptible to the malady." Beers points out that though the condition of zoophilpsychosis was eventually dropped, adversaries of the animal movement retained distinct archetypes of the animal rights "bogeyman" who was framed as a fanatic, antihuman, a national traitor, anti-progress, and/ or mentally ill. These "bogeymen" kept being deployed because they were politically useful and curtailed the effectiveness of critique, as Beers writes, they worked with "the entrenched belief systems that subjugated other species to humans and women to men" (2006, 25).

The bogeymen and mad women stereotypes did not simply revolve around gender but also racialized assumptions about civilization and ideas of progress. Supposed zoophilists became bogeymen because they were out of step with "civilized communities," according to Dana (382). This view supposed that "civilized" communities had made improvements on past practices. Here Dana's depiction of the wrongness of advocates' feelings worked to establish the rightness of contemporary practices. In other words, the zoophilists suffered not only disproportionate feelings but also feelings that were anachronistic (out of date) and also out of place (outside of civilized communities). Other versions of this ploy persist. Animaladies may be acknowledged in one context (the past, or other cultures/races/communities), while being refused in the present or in our own communities. We see this in the reassurances offered by animal welfare approaches in animal agriculture; improvements made to address a dysfunction (in care, in housing, in pain management) are likely to be acknowledged retrospectively and only with the caveat of perpetual improvement. Nekeisha Alexis' chapter comparing pro-slavery writings and humane farming narratives also shows that such apologia relies on racist and anthropocentric illusions of progress. We see this when animal advocates fixate on the activities of specific communities and their animal practices, while not acknowledging the ways that such a focus works through an optic of racism (Kim 2015). Animaladies are therefore not the same across all contexts; indeed, the identification of them as animaladies depends very much on the social and political contexts in which they occur.

Animaladies are racialized, gendered, and class-specific, and they can be repeated uncritically by animal advocates and scholars too. Traditional theorists of the animal rights movement also, though perhaps unwittingly, drew on animaladies as excessively gendered and overly emotional—they emphatically denied any connection to sentimental, little old ladies in tennis shoes. In Peter Singer's preface to the 1975 edition of *Animal Liberation* he is at pains to show that his thinking was motivated by reason: "We didn't 'love' animals. We simply wanted them treated as independent sentient beings that they are, and not as a means to human ends" (1990, 2). Tom Regan also stresses the rationality of *The Case for Animal Rights* by arguing that he intended it to "give the lie, once and for all, to all those opponents of animal rights who picture everyone in the movement as strange, silly, overly emotional, irrational, uninformed, and illogical" (1983, 94). Instead, Regan argues, those who believe that about the movement are themselves "strange, silly, overly emotional, irrational, uninformed and illogical" (1983).

Feminists by and large could not reject their critics in the same way that Regan does, with a repudiation that affirms rather than undoes the insult. A feminist approach to the imbrication of emotions and reason, the irresistible entanglement of rationality with feeling (see Gruen's (2015) entangled empathy and her chapter here), sees such separation as patriarchal, illusory, and undesirable to maintain. To separate emotion from reason once and for all is, from a feminist ethics of care perspective, akin to reinstating a mind/body dualism, a false dichotomy that underscores not only the masculine/feminine binary but also the human/animal one (see Adams 1990). The risk of being seen as emotional, and being seen to be emotional about *animals*, poses a particular kind of risk that is both gendered and species-specific. And even women who do write with their animals in mind are, as Susan McHugh explains, frequently doing so *apologetically*, apologizing for taking their cross-species friendships seriously (2012).

Animal Studies scholarship is frequently marked by the same sorts of anxieties over legitimacy. In part this might relate to resistance to alliance with other political movements. Kymlicka and Donaldson have pointed out that progressive social movements have largely failed to recognize animal rights as a social justice issue, making the animal movement an "orphan of the left." They argue that these other movements fear that focusing on animals will "erode the moral seriousness with which human injustices are treated" (2013, 4), and they identify the fear of displacement, trivialization, and the "depth of our cultural inheritance" as major reasons for resistance to the animal movement as a whole. But there is also another possible element at work here, which is the fear that thinking about animals not only trivializes human injustices, but that it is fundamentally unreasonable, "crazy." Instead of being the "orphan of the left," animal advocates are perhaps more readily identifiable as the left's "crazy aunt," given the predominance of women in the animal movement (Munro: 2001). We might say that Animal Studies has, so far, done all the right things according to "proper" academic inquiry, but perhaps we have neglected to take the matter of madness itself more seriously.

When Donna Haraway writes about having "gone to the dogs" (2003, 5) she too is acknowledging the risk of being seen as having degraded her work, losing a form of respectability associated with "proper" (read: human oriented) academic work. She is also indicating her allegiance to the dogs and signaling a familiar feminist trajectory: redefining an insult as a compliment, reclaiming what is repudiated as a badge of pride. This tactic is familiar, appearing recently in the "pussyhat project" in the United States (to protest Donald Trump's sexism), the group Mad Pride (Lewis 2010) protesting the medicalization of mental difference, the "mad fucking witches" in Australia, and in a similar vein, Yvette Watt's *Duck Lake Project* (discussed in Chapter 13): all are political campaigns that return the gaze and resituate an animalady as an act of defiance. Reclaiming the "crazy" is an important tactic, as Brittany Cooper describes in *Eloquent Rage*: "Talking crazy" is either a "compliment or indictment" and indicates a "flirtation with ideas that skirted the line between being profound and absolutely nonsensical" (2018,11). "Crazy" projects reclaim a label, reverse the gaze, and make a great deal of sense, but they are risky. None of them are in any simple way a celebration or trivialization of the distress that "madness" or "craziness" may involve. Rather they are attuned to the risk associated with being seen as mad and as angry, as unable to adjust to injustice, like the character of Yeong-hye in Han Kang's *The Vegetarian*, described by Hayley Singer in her chapter, as unable to "go docilely into quarantine" nor able to be "re-adjusted to fit a carnist perspective of life."

Animal activists and scholars often readily accept the risk of being derided as misanthropes who have become incomprehensible: neither properly gendered nor properly human. From twentieth-century "zoophilpsychosis" to twenty-first-century "species identity disorder" and "orthorexia" (discussed in Chapter 9), the disciplining of relations between humans and animals in terms of norms pertaining to gender, race, and "humanness" persists.

We've divided the chapters that follow into three interrelated themes: *dismember, disability,* and *dysfunction.* We'll leave the pleasures of discovery and interpretation to you, but we do want to say a bit about our decisions.

One of the many ways that animaladies function is to alienate, to divide, to, in a word, *dismember.* Of course, animals are routinely, literally dismembered for food and after (and sometimes during) laboratory experiments. Animals are also dis/membered when they have their infants removed from them or when their social groups are altered by human choices and actions. Scholars and activists who are working to reveal the implications of these harms face particular kinds of dangers, including the psychological distress and trauma brought on by witnessing the violence enacted on animals in slaughterhouses, hunting grounds, or laboratories. These scholars/activists allow themselves to be moved by the knowledge generated in painful relationships, to hold on to the violation they are attending to while holding off the cultural and political mechanisms that seek to sanitize and normalize dismemberment. The mechanization and distribution of the process of killing through many hands produces a "double dysjunction," as Noele Vialles describes it, where "we are left without any 'real' killing at all" (1994, 159). If these dismemberments cannot be attributed to any one actor, it starts to seem like they do not happen, and the emotions and knowledge these nonhappenings evoke have no place either. The scholar in the slaughterhouse or in the laboratory and the activist at the hunting ground or at the stockyard witness events that "do not happen" for audiences who may not care to know (Probyn-Rapsey 2013), and thus their grief is "misdirected" or beyond sensibility (see Chapter 5). Cloaking emotion is another way to maintain the reason/emotion, mind/ body divide. Allowing the sadness, the anguish, the pain to be seen and heard and felt, although risky in various ways, helps us to re-member. Through art, the discomfort of dismemberment can become real again, as we see in lynn mowson's work on the cover of this volume and in the chapter she has written for this volume.

When it comes to the animaladies associated with the framing of *disability*, Sunaura Taylor's book *Beasts of Burden* is central (2017). Taylor explores the shared histories of animalization that have affected those with bodies and minds seen as "lacking" or deficient: animals (other than human) and those humans with physical or cognitive disabilities. Taylor resists the impulse to see the comparison between animal and disabled person as pejorative, instead thinking of it as a preface to an alliance, a potentiality that has been strained by those within animal advocacy as much as outside of it. In her book, Taylor sharply focuses on Peter Singer's dismissal of severely cognitively disabled humans, as does Alice Crary in her chapter in this volume. Some Animal Studies scholars have used mental illness and cognitive disabilities as a rhetorical weapon in their criticisms of speciesism. Guy Scotton outlines how Animal Studies scholars have been guilty of labeling opponents "mad" and "crazy" too, perpetuating the problem of stigmatizing mental illness and cognitive difference. Physical disabilities, cognitive disability, and psychological disorders are a particularly important set of animaladies to consider as potential sites for alliance, as Hannah Monroe's chapter on autism and interspecies friendship shows and Nik Taylor and Heather Fraser's analysis of women, companion species, and anxieties highlights.

Animaladies cautions against an exclusive emphasis on the socially constructed nature of madness; to be seen as "mad" is one thing, but to experience madness as profound psychological distress is quite another, and not one that is likely remedied only by attributing it to the discontents of social life in general. As Jane Ussher reminds us in relation to the "mad woman," we must also listen to what she has to say rather than speak for her as an emblem of heroic feminist resistance. As Nik Taylor and Heather Fraser's chapter shows, the psychological needs and distress of participants in their study of human–animal comfort are palpable and resistant to generalization. The same applies to the animals—alleviating their distress brought about by violence on factory farms and cruelty in laboratories is not only a matter of identifying the architecture of power that imprisons them but must also consider the actual lives they lead as individuals, the freedoms and pleasures they deserve after the horrible pains they endure. As many of the chapters in this book argue, we cannot in any simple way embrace "madness" wholeheartedly, whatever that might look like. But we can, perhaps more modestly, understand how it is used to preemptively frame animal advocacy and Animal Studies as far from the norm while also stigmatizing dis-ease, mental difference, neurodiversity, discontent, and disability.

The chapters in the section called *dysfunction* draw on the myriad ways in which relations between animals and humans are made dysfunctional or depicted as such. Another "dis" or "dys" word, it describes behaviors or functions that are broken, disturbed, disrupted. Again, our focus here is not on labeling the good as opposed to the bad, but in examining how easy it is for some behaviors (but not others) to be depicted as "the problem" (like the "crazy cat lady," the "hysterical" vegan, or "crazy" activist) while letting other damaging relations (such as in factory farming, humane farming, duck shooting) off the hook. Those who bear the brunt of these damaging relations are also demonized, or outcast with their animals as Liz Bowen's discussion of precarious outcasts shows, and pattrice jones and Cheryl Wylie's play on dammed and damned feelings illuminates. These emotions are a source of strength too, as Katie Gillespie asks at the end of her chapter, "Who would we be without this madness, this feeling, these politicized manifestations of witnessing and care?" Together, these three sections and the three words—"dismember," "disability," and "dysfunction"—account for the various formations of animaladies: those damaged relations that feminist Animal Studies acknowledges, some to be reclaimed in the spirit of defiance and others disclaimed as acts of violence that need to end.

References

Adams, C. J. (1990), *The Sexual Politics of Meat: A Feminist-Vegetarian Critical Theory*, New York: Continuum.

Ahmed, S. (2017), *Living a Feminist Life*, Durham: Duke University Press.

Beers, D. (2006), *For the Prevention of Cruelty: The History and Legacy of Animal Rights Activism in the United States*, Athens: Swallow Press.

Chesler, P. (1972), *Women and Madness*, New York: Doubleday.

Cooper, B. (2018), *Eloquent Rage: A Black Feminist Shares Her Rage*, New York: St Martin's Press.

Dana, C. (1909), "The Zoophilpsychosis: A Modern Malady with Illustrative Cases," *Medical Record: A Weekly Journal of Medicine and Surgery*, 75: 10.

Dickinson, E. ([1896] 1996), "I felt a funeral in my Brain," *The Essential Emily Dickinson*, Selected by Joyce Carol Oates, New York: HarperCollins, 12–13.

Donaldson, S. and Kymlicka, W. (2013), *Zoopolis: A Political Theory of Animal Rights*, New York: Oxford University Press.

Gilbert, S. M. and Gubar, S. (2000), *The Madwoman in the Attic: The Woman Writer and the Nineteenth-Century Literary Imagination*. New Haven, CT: Yale University Press.

Gruen, L. (2015), *Entangled Empathy*, New York: Lantern Press.

Haraway, D. (2003), *Companion Species Manifesto: Dogs, People and Significant Otherness*, Chicago: Prickly Paradigm Press.

Kim, C. (2015), *Dangerous Crossing: Race, Species and Nature in a Multicultural Age*, New York: Cambridge University Press.

Lewis, B. (2010), "A Mad Fight: Psychiatry and Disability Activism," in L. J. Davis (ed.), *The Disability Studies Reader*, 339–352, New York, Routledge.

McHugh, S. (2012), "Bitch, Bitch, Bitch: Personal Criticism, Feminist Theory and Dog-Writing," *Hypatia* 273: 616–635.

Munro, L. (2001), "Caring about Blood, Flesh, and Pain: Women's Standing in the Animal Protection Movement," *Society and Animals* 9 (1): 43–61.

Probyn-Rapsey, F. (2013), "Stunning Australia," *Humanimalia: A Journal of Human/Animal Interface Studies* 4 (2): 84–100. Available online, (accessed March 2, 2018).

Probyn-Rapsey, F. (2014), "Multispecies Mourning: Thom van Dooren's Flight Ways: Life and Loss at the Edge of Extinction," *Animal Studies Journal* 3 (2): 4–16. Available online: http://ro.uow.edu.au/cgi/viewcontent.cgi?article=1116&context=asj (accessed March 2, 2018).

Regan, T. (1983), *The Case for Animal Rights*, Berkeley, CA: University of California Press.

Russell, D. (1995), *Women, Madness, and Medicine*, Cambridge: Polity Press.

Showalter, E. (1985), *The Female Malady: Women, Madness, and English Culture, 1830–1980*, New York: Pantheon.

Singer, P. (1990), *Animal Liberation*, New York: Avon Books.

Taylor, S. (2017), *Beasts of Burden: Animal and Disability Liberation*, New York: New Press.

Ussher, J. (1991), *Women's Madness: Misogyny or Mental Illness?* Amherst: University of Massachusetts Press.

Vialles, N. (1994), *Animal to Edible*, Cambridge: Cambridge University Press.

Part I

DISMEMBER

Chapter 1

JUST SAY NO TO LOBOTOMY

Lori Gruen

The case of lobotomy, despite its extremity, provides a poignant illustration of the precarity or threat of affect (or emotion or sentiment or care) in our relationships with each other and other animals. In this chapter, I will explore the ways "affect" has been understood, the way it has been feminized, rejected, and policed, and interrogate attempts to have it removed from our discourse and theory, from our activism, even from our brains! These efforts are inevitably unsuccessful, but the impact of the rejection of affect has nonetheless been intense and I think it has hindered collective efforts at making the world better for animals, or at least less horrible. Indeed, we are not able to fully be in meaningful relationships when a big part of our cognitive/affective capacity is cut off, even if it's not technically "cut out."

I will first discuss this scientific malady of remarkable proportions, the lobotomy, and then I will turn to an analysis of the rejection of affect, and finally I will revisit the sort of affect I have been advocating, what I call entangled empathy (Gruen 2015), which resists the division between reason and emotion and seeks to not just enhance our pursuit of justice but provide us with meaningful, caring "crazy"—in the sense of counter-normative, excessive ways to enrich our relationships in that pursuit.[1]

Lobotomies

Walter Freeman was an "overzealous" showman who traveled the United States performing and promoting lobotomies. He is reported to have performed over 2,500 lobotomies on patients from twenty-three states. His first patient, in 1936, was a housewife from Kansas, the first of many more housewives who had parts of their brains ablated. Another housewife, a 29-year-old, was the first to receive the transorbital lobotomy. His last lobotomy was performed on another housewife, Helen Mortensen, who was subjected to her third lobotomy and died from a brain hemorrhage. That was in February 1967 and it was after her death that Freeman was banned from operating.[2]

In 2005, National Public Radio in the United States did a story on one of the youngest people to have a lobotomy, Howard Dully, who was just twelve years old (NPR 2005). The story is shocking, and Dully (with Charles Fleming) wrote a riveting memoir about his experience (2008). The NPR story included a few oral histories, one was from a housewife, Patricia, who seemed glad to have had a lobotomy as she was suffering from depression. Her husband had arranged for her to have the lobotomy and apparently this interview was the first they had talked about it (the lobotomy was in 1962—forty-three years earlier). In that interview Patricia says she doesn't remember anything at all about it and Glenn says, "We were coming back from San Jose, following the operation, and Pat informed me that she couldn't wait to get home because she wanted to go down and file for divorce." Patricia responds, "Hmmm … Don't remember that at all. I don't think I said it." Glenn says, "I think I just went on driving and ignored the situation and began to wonder to myself, 'How much good did this operation accomplish?' Really, I can see no changes in most areas except she's much easier to get along with." Pat says, "I was a more free person after I'd had it. Just not to be so concerned about things … I just, I went home and started living, I guess is the best I can say—just started living again and was able to get back into taking care of things and cooking and shopping and that kind of thing."

The idea of altering the brain in order to facilitate social compliance and eliminate undesirable behavior has an interesting history. Surgical interventions reached their heyday in the 1940s as mental asylums were brimming over with cases after the Second World War. In that decade alone, more than 18,000 lobotomies were performed in the United States and tens of thousands more in other countries.

In "Psychosurgery: Intelligence, Emotion and Social Behavior Following Prefrontal Lobotomy for Mental Disorders," Freeman wrote:

> [T]he frontal lobes are essential for satisfactory social adaptation … certain individuals may suffer from perverted activity of these areas and may become capable of better adaptation when these lobes are partially inactivated … Partial separation of the frontal lobes from the rest of the brain resulted in reduction of disagreeable self-consciousness, abolition of obsessive thinking, and satisfaction with performance, even though the performance is inferior in quality. The emotional nucleus of the psychosis is removed, the sting of the disorder, is drawn. (1942, vii)

He goes on to describe the history of surgical intervention in mental disorder, since female genitalia were often thought to be the site of the disorder. He noted that

> surgeons began searching about here and there for offending organs that were supposed to be causing mental disorder. The first attack … was upon the internal genitalia, particularly the ovaries, because of the notion that functional nervous disorders were produced by the wanderings of the uterus

into various parts of the body ... After several years of experimentation and the sacrifice of many thousands of ovaries, the conclusion was reached that castration in the female, while sometimes exercising a temporarily beneficial effect, was not the solution to the problem of mental disorder. On the contrary, it frequently brought in its wake a train of undesirable symptoms that increased the discontent and often the misbehavior of the sufferer. (1942, 5)

In the early days of medicine, when all sorts of different theories were floated about the causes of illness, it was not uncommon to think that they were due, in some way, to "a depraved state of the humors" which was usually the fault of the sufferer. Thomas Sydenham, a well-known seventeenth-century British physician, noted that female hysteria was the second most common disease, just behind fevers. He attributed the malady to "irregular motions of the animal spirits," which were caused by "some great commotion of mind, occasioned by some sudden fit, either of anger, grief, terror or like passions."[3]

Two hundred years later, it's only slightly surprising that women were the most common patients—and even today, women suffer more mental disturbances than men. A short piece in *Psychology Today* reports that even though it may be simple to attribute the "epidemic of mental illness among women" to hormones, or the idea that women are more emotional, "The truth, though, is that psychiatrists aren't really sure why mental illness is more common among women" (Young 2015). Perhaps the reason this remains mysterious is because there still are more men than women asking the question.

One male psychiatrist, Gottlieb Burckhardt, was the first person on record to perform a surgical procedure to relieve the commotion of the mind. Burckhardt was a superintendent at a small Swiss mental hospital, and in 1888 he was the first to experiment with removing or destroying parts of the brain to address "perverted" problems of the mind. He experimented on six people. His work was later criticized because he was unaware of brain functionality and, in fact, was not trained as a surgeon. One of the patients died five days after the operation from epileptic convulsions, one "improved" but later committed suicide, another two showed no change, and the last two patients became "quieter."

About fifty years later, brain surgery for behavioral maladies got going again after a paper presented at a neurological conference in London renewed excitement about surgical intervention for mental disorder. At that conference, John Fulton and his graduate student Carlyle Jacobsen from Yale reported on an experiment they performed on two young chimpanzees, Lucy and Becky. The use of chimpanzees in the history of lobotomies is often overlooked, but when it is mentioned, it is usually just these two chimpanzees that are noted. The full impact of Fulton's research on chimpanzees is largely absent. Indeed, at the conference, Fulton and Jacobsen only mentioned the two young chimpanzees who survived the procedure. Ross, another young chimpanzee, died of dysentery while in Fulton's lab. Infant chimpanzee Lu died from meningitis following frontal lobe extirpation. She was involved, with Lucy, Becky, and another chimpanzee, Jerry, in Fulton's laboratory where they were trying "to create a group or colony of experimental defectives" (R.M.Yerkes Papers). Jerry died following an experimental operation, Lucy was killed and her brain prepared for study. An annual report from the lab in 1937 states that "Becky will be sacrificed within a few months" (R.M. Yerkes Papers).

At the London Conference, in their report to the attendees, Fulton and Jacobsen describe the behavioral change that occurred after they ablated the frontal lobes of Lucy and Becky's brains. It is a small sample, just two chimpanzees since the others didn't survive, and importantly, the behaviors they reported were opposite one another. One became more agitated about things she didn't care much about before, the other less agitated about things she did care about. This should not have led anyone to the conclusion that this procedure be attempted on humans, and Fulton allegedly thought as much (although he continued with his lobotomy work on chimpanzees). In addition, in the 1930s, very little was actually understood about chimpanzee behavior, so what the experimenters report they believed about behavior must be viewed with skepticism. The first infant bred and born in a laboratory, Alpha, was in 1930 in part to begin to understand chimpanzee development and behavior. Altering the developing brain of infant and young chimpanzees, without understanding their species' typical behaviors and their individual personalities, could not lead to any respectable conclusions. Sadly, this sort of scientific rationality was a large part of early biomedical research, and in the case of chimpanzees, at least,[4] this "foolishness" continued for another fifty years.

At the conference, two important figures in the development of human lobotomies were in the audience, Egas Moniz and Freeman, who within months developed techniques and began performing lobotomies on humans. The desire to destroy unwanted emotion actually overtook reasoned assessment ironically in the name of scientific rationality. Lobotomy's reputation once ran so high that the Nobel Committee awarded the prize in Medicine and Physiology to Moniz in 1949. But within the next decade, lobotomy fell out of favor and its memory increasingly was vilified.

The operation's descent into disgrace had many causes. For one thing, lobotomy never had a scientific basis and animal work wasn't thought to be useful as a foundation for the procedure because researchers couldn't admit that animal models of mental illness existed. That was a barrier that the infamous Harry Harlow most notably sought to break with his psychological experiments on infant monkeys (Gluck 2016). Similar experiments on sensory and maternal deprivation were also being performed on chimpanzees at that time in an attempt to create animal models for despair, depression, schizophrenia, anxiety, addiction, and the like (R.M. Yerkes Papers). Of course, the advent of psycho-pharmaceuticals also played a role in ending the lobotomy. In the history of medicine, the lobotomy is one of the rare examples in which those within the biomedical community themselves share the negative assessment.

In 2005, Dr. Nuland, for example, began a discussion of the history of lobotomies noting:

> Major steps in scientific progress are sometimes followed closely by outbursts of foolishness. New discoveries have a way of exciting the imagination of the well-meaning and misguided, who see theoretical potentialities in new knowledge that may prove impossible to attain. On occasion, the seemingly imminent is later shown to be far further off than originally thought, yet still possible to achieve. More frequently, the apparent prospect is revealed to be the result of unrealistic hypotheses based more on wishful thinking than on fact. In no branch of human thought have erroneous leaps of this kind been more prevalent than in that peculiar mix of science and art that goes by the name of medicine. (2005)

Nuland had a personal interest in the case of lobotomies, as someone who suffered from depression he was hospitalized and almost had a lobotomy, but a young physician recommended electric shock therapy instead. He ended up having a long notable career. Of this sorry chapter in medical history, Nuland writes: "Freeman's foolishness influenced an entire generation of psychiatrists, neurologists, and neurosurgeons and devastated the lives of tens of thousands of patients and their families." This tragic path "can serve as a cautionary tale about those who come to believe that they are beyond the restraints of judicious professional behavior."

Despite the demise of the surgical lobotomy and the scientific maladies that followed, there is a more ideological sort of lobotomy—a cutting off of feeling or affect that has been much more successful, indeed entrenched, and it is to this sort of cognitive alienation that I now turn.

Affect

In the United States in recent years, due in large part to the wide availability of social media, the murder of black people by police officers has come to wider attention.[5] One of those killings, of Philando Castile, was live streamed on Facebook by his partner Diamond Reynolds, whose four-year-old was also in the car. The video showed the police officer with his gun drawn on Castile who is in the driver's seat bleeding out.

Many people who watched the video commented on Reynold's composure. She was calm and appeared quite rational. For both black and white commentators, her composure seemed a sign of her trustworthiness as a witness. Of course, she and her young daughter needed to make it out alive, so there was good reason for her to be calm in the face of a panicked cop with a gun. But as Melvin Rogers notes, there is something else going on here:

> We would never expect others to display such composure in the face of such traumatic circumstances. We would not penalize their failure of self-control by tying it to untrustworthiness. In fact, we think, and rightly, that emotional eruptions at precisely this moment are appropriate ... It makes perfect sense to come undone in that moment ... And yet, we find ourselves holding Reynolds and other black folks, often women, to a higher standard as a prerequisite to be considered trustworthy, capable of accurately recounting the injustice that has just been committed against them. In doing so, we commit another form of violence, the reverberations of which most assuredly affects the mental health of black folks, reminding us, yet again, that what is expected of Black Americans is not expected of whites. It is demanded that we hold in and contain what should rightly be released: screams and tears. In short, pain. (2016)

I don't know if it's true that "we" don't hold "others" to this standard, but that is not really the point. This is certainly violence, what Dinesh Wadiwel (2015) and others have called epistemic violence. In particular, what Rogers is decrying is a type of testimonial injustice that denies an individual credibility in virtue of prejudice on the part of the hearer. Most women and people of color suffer from this form of injustice and thus experience all manner of credibility deficits. When exposed to long-term epistemic injustice, many will alter their/ our mind-sets and behaviors in an attempt to conform to the epistemic norms, even though their/our attempts will seldom succeed. Most of us are familiar with the experience of having our perfectly reasonable claims dismissed because of who we are (perhaps in addition to a disagreement with what is said as well as how it is said). There is a growing and important literature on the ways these injustices further distort our relations with one another, and I think it is particularly important to be mindful of these credibility deficits in Animal Studies when we encounter scholars who express skepticism about what we might be claiming about what we know of other animals' experiences, for example. There is a cognitive alienation—a type of lobotomy—at work here.

So-called experts have been quick to diagnose Reynold's calm response. Jim Hopper, a psychology instructor at Harvard Medical School, said her behavior was consistent with a dissociative state. "In the immediate aftermath of horrific violence, he said, victims don't always sob." "People are literally not feeling in their body what's going on," Hopper said. "That circuitry can basically shut down" (2016).[6]

This shutdown is emotional distancing—cutting off one's affect—and it is to some extent a "flight" reflex in the face of extreme horror, but I also think it is important to explore the social/political ramifications of the view that emotion must be cut off in an attempt to muster some credibility resources. In Reynold's case, she at least had good reason to believe that her chance of surviving would be greater if she did not display emotion. I see this as a coerced affect lobotomy, and it is much more widespread than the surgical lobotomies I discussed earlier.

There has been a lot of interesting, and conflicting, work on affect theory over the last decade. Some understand affect as the automatic, reflective phenomena that can be explained through forms of scientific reductionism reinvented as "embodiment." Others see the affective turn as a measured response to the rejection of sentiment, particularly when in literature it is deployed to alienate and subordinate. As Colin Dayan recounts:

> [T]he language of sentiment animates subordination. A slave, a piece of property, a black cat—once loved in the proper domestic setting, they arouse a surfeit of devotion, bonds of dependence that slavery apologists claimed could never be felt by equal ... Narratives of humane care are always conducted by the free in the name of the bound, their emotive impulse turning away from political action and toward ... the neoliberal "performance of pathos." (2015)

Against this take, the affective turn is an intervention in political/literary invocations of domination. In philosophy, affect or sentiment has stood in opposition to reason, in judgments of value as well as in perceptions of good and bad, right and wrong.

This opposition between reason and emotion is at the heart of why animal ethicists, like Singer, make arguments based on our capacities to reason about similarities. It is in virtue of other animals being like us in "morally relevant" ways that we owe them consideration. We come to this through reason not sentiment. Recently, J.M. Coetzee, in a lecture about animals, echoed an early sentiment expressed by Singer, exclaiming, "I am not an animal lover ... Animals don't need my love ... I don't care about love. I care about justice."[7]

I've been asking why not both? Coetzee is admitting he cares—just not about love? Why? The answer, of course, is political and gendered and I think it is based on a misperception of the structure of our cognition.

Singer, Coetzee, and so many others are in love with a liberal fiction that our relationships of care are "private," beyond the scope of ethical reason, and optional. Of course, most people don't want the State too intimately involved in their lives, but the continued reliance on a strict public/private division is not only a faulty notion, but it also perpetuates gendered power dynamics. White men have a sense of themselves that is shaped by imagining themselves primarily as thinkers and actors, something they admittedly learn in relationships with others, but their ethical/political identity kicks in once they are beyond those relationships, and they imagine themselves as post-relational.

According to this post-relational fiction, care, love, and sentiment are private, feminine experiences; justice on the other hand is seen as masculine and unsentimental. Of course, this binary thinking not only builds on stereotypical gender roles that preclude the idea that men are caring and obscures gender-queer expression, but it also ignores the particularity of caring relationships, which are informed by racial, economic, ethnic, cultural, and differently gendered experiences, that are fundamentally about justice or, more precisely, injustice. Perhaps most importantly, those of us who are not under the illusion that one can ever be post-relational and who find that fiction undesirable to start with are further marginalized when we reveal that this view is untenable. We are denied credibility, brushed off—we are subject to testimonial injustices.

Justice and love aren't at odds because they can't be disentangled in any meaningful way; they are mutually informing. Any compelling moral or political theory has to attend to the interconnections between cognition/reason and affect/emotion. Rather than generating distance between us and them, justice and love, in other words, to avoid this sort of lobotomy, we need courage, even while being marginalized, to bridge perceived gaps between reason/emotion and self/other by recognizing the ways that each side informs the other without collapsing into it.

Entangled Empathy

I've been thinking about building a bridge for many years and have developed the notion of entangled empathy with the hope that it will be strong enough to span these perceived gaps. It is a view that involves refining one's perception to more accurately see the concerns of particular, situated others; to recognize the ways we are in relationships that entail responsibilities to ourselves and others; and to do our best to make these relationships go better. Entangled empathy is a type of caring or loving perception that inevitably blends emotion and cognition.

While entangled empathy is part of the feminist ethics of care tradition in animal ethics, I was also inspired by the work of Iris Murdoch (1970). Murdoch is critical of what I'm calling this post-relational moral agent who is conceived as someone who already has a handle on any given situation, who knows what he is doing, whose thoughts and intentions are "directed towards definite overt issues," and whose responsibility is a function of impersonal knowledge. I love the irony in Murdoch's comparison of this sort of moral action with shopping, which is usually feminized—the agent enters the shop "in a condition of totally responsible freedom" and surveys the products and chooses to purchase one product or another. Shopping is public, and one understands the shopper through his choices and actions, much as the modern economist understands the preference satisfaction of economic man through his choices and actions. The inner life of the shopper, of economic man, and of the moral agent understood through modern moral theory is "parasitic" on his outward behaviors.[8] But there really is only so much that one sees of that "inner life" through outer behavior, it is an anemic view of our inner lives, if it provides anything at all.

Indeed that we perform behaviors that are meant to project meaning very different from what we internally experience is a good part of what therapy and other efforts at self-reflection start with. These outward behaviors become the focal point of ethical choice and action, and the inner life of the agent remains mysterious or, when accessible, thought to be beside the point of ethics.

Entangled empathy is, in part, trying to solve that mystery and make the embedded and embodied process of moral perception more accessible. I've argued that entangled empathy involves movement from the first personal to the third personal points of views. To meaningfully attend to the particularities of another being's interests, desires, vulnerabilities, etc., as they occur in particular contexts, under certain norms, within social relations of all sorts, and to avoid projecting one's own state of mind on others, this movement from me to you and back, through these shifting perspectives, is key. This process allows one to check oneself to avoid projection and to correct misperception.

Entangled empathy isn't simply a form of ethical attention but a particular form of caring or loving attention, attention that is directed toward another's flourishing. And, of course, people flourish in different ways. Through entangled empathy we see not only who we ought to attend to and what their situation evokes but also whether and how we should respond in ways that help to preserve and promote their well-being.

We can and do go wrong. One might err when they fail to grasp cultural understandings of the values and disvalues that are presupposed by empathetic experience. Evaluating those presuppositions is one of the tasks for getting empathy right. An empathetic agent's evaluations, deliberations, and choices are shaped, at least in part, by the social context in which he or she lives. Various social institutions and norms affect expectations. Empathetic engagement will often involve examining the conditions under which both the empathizer and the one with whom she is empathizing form their desires and come to appreciate their particularity. Entangled empathizers will be attuned to the distribution, both just and unjust, of hermeneutical resources that have an impact on desire and action. An astute empathizer will thus try to answer a host of questions, informed by the particular situation they are in, including: What predispositions do I have, and does she have, and how do they affect our different levels of confidence in formulating our desires and acting on them? How do cultural understandings of value impact our sense of our worth? Were certain paths (emotional, deliberative, and/or material) closed to me or to her by familial, educational, social, political, racial, gendered, economic, or religious barriers or prohibitions? Will my empathic success or failure depend on the social position I occupy? The social position she occupies? Do people like me culturally or historically have less chance of empathically succeeding and of "making a difference"? Are there any resources available or that I can cultivate to more successfully empathize?

I imagine a slightly different set of questions when the relationship is with nonhuman animals. What were the early rearing conditions this animal experienced and did that shape her current experiences? Chimpanzees who were singly housed and nursery reared have very different ways of being than those who had more socially appropriate upbringings. What sort of species-typical behaviors does this kind of being usually engage in and does she have opportunities to engage in those behaviors? For example, is there enough space to do what they normally would do? What sorts of social relationships are important, whether they be with conspecifics or animals from other species, including humans? Is this animal able to be alone, if she chooses? Is she able to make choices about who to spend time with, where to be, what and when to eat? Are these the sort of choices that are meaningful to this particular animal?

Answers to these questions, and others, will require us to develop better skills at perceiving and noticing the complexity of moral experiences in the world allowing us to navigate the "values round about us" as richly and fully as we can. It is a very different set of choices that we are considering than those enacted through typical transactional experiences, alluded to by Murdoch, in which mature actors are thought to be rational economic agents, expressing value through choices.

These "rational" actors, those who are lobotomized, are not able to engage empathetically and that means that so many are not attending to the sorry state of emergency that we live in. Of course, most of us are busy and absorbed in our own problems, projects, and plans, and many of us also have the luxury of not thinking about the significant problems that, for example, black people in the United States, Indigenous people in Australia, most people around the globe, and the majority of other animals are confronted with, in one form or another, almost every day. Those of us with relative privilege do not need to worry about being arrested, assaulted, profiled, starved, or shot in the streets. But we can pay attention and empathize with all those who are disregarded and do something to stop it, even though it is difficult, often painful, and we may be ostracized, ridiculed, or pathologized in our solidarity with others. This is necessary and possible, since the empathetic, caring, side of our cognitive architecture has not, fortunately, been surgically removed.

Some theorists are suggesting that we ought to remain figuratively lobotomized, although not in those words, and that empathy is a problem. Paul Bloom is probably the most vocal critic of empathy. He worries that we will end up caring about the wrong things, the wrong people, we'll be too parochial, we'll be subject to all sorts of biases (2014, 2016). I have written a bit about this, but I think these worries amount to a fairly standard worry about emotion being unhinged and unchanging.

There are different ways to understand what emotions are, and while Bloom seems to be thinking of emotion as simple sensations or feelings, emotions can be explained by the reasons that give rise to them and can be altered in light of those reasons. Emotional states can also alter reasons. Emotions are the sorts of things that we can be taught or conditioned to feel or not feel and reasons can be changed based on our affect. But, importantly, entangled empathy is not itself an emotion, but a *process* or *method* of moral perception.

I want to end by clarifying what I have in mind by the entanglement part of entangled empathy. I have a relational conception of the self and argue that our agency is co-constituted by our social and material entanglements. Social entanglements often extend beyond the human and far beyond our geographical location. By material entanglements, I have in mind the sense of materialism that is linked to our socioeconomic opportunities and limitations. I am also thinking about the new sense of materialism, which would include our entanglement with the food we have access to, the safety of our physical environment (e.g., water, air, particulate matter, toxic exposure), and the nature of our microbiome. These entanglements are quite complex and include our relations to the child slaves who harvest cocoa for chocolate; the orangutans who are on the brink of extinction due to our consumption of palm oil and palm products; those working in sweatshops who provide cheap clothing; our greenhouse gas emitting choices for food that are creating climate refugees. All of these actions, in part, constitute who we are.

Our identities are not simply "socially constructed," rather we are who we are at any particular time as an expression of entanglements in multiple relations across space, species, and substance. I am here inspired by the notion of entanglement from feminist philosopher of physics Karen Barad, who suggests that:

> Matters of fact, matters of concern, and matters of care are shot through with one another ... Ethics is therefore not about right responses to a radically exteriorized other, but about responsibility and accountability for the lively relationalities of becoming, of which we are a part. Ethics is about mattering, about taking account of the entangled materializations of which we are part, including new configurations, new subjectivities, new possibilities. (Barad 2012)

We should care about others because they are fundamentally part of our own agency. My failure to respond to others is not just a failure of my reasoning and my affect, but I may come to see it as a weakness in who I am or an error in my sense of agency.

Of course, it would be arrogant or naïve to think even though we co-constitute each other's agency that we can really ever, truly understand another. Sometimes, especially when it comes to animals, those of us in particularly intimate relations think we do really know. I would caution more humility here. But I worry that too often we take the possibility that we can't fully understand as an excuse to not to even try to take the perspective of another. The clearest case of this lately is that white people in a culture of anti-black racism cannot understand the full weight of years that burden those who experience racism, as well as the feelings of invisibility, rejection, extrajudicial violence, and disrespect that result from it. For some this has meant that white people need to step away from the world; for others it has meant that we need to work to develop and deepen our empathy.

Entangled empathetic moral attention involves working through complicated processes of understanding one another and other animals in situations of differential social, political, and species-based power. Usually what we "get" is just a glimpse. That we never really "know" cannot be an excuse to opt out of working at it. I take this to be a failure of both imagination and agency. To avoid this failure, we need to *say no to lobotomy* and deepen our care for each other, for other animals, as well as for ourselves.

Acknowledgments

I would like to thank Fiona Probyn-Rapsey for conceiving and planning the wonderful conference at which an early version of this chapter was presented and to Dinesh Wadiwel and HARN for bringing me to Sydney. The audience at Animaladies was terrifically receptive and I am grateful for and was inspired by that. pattrice jones helped me come up with the title of this chapter. My deepest gratitude is to Njeri Thande. Without her, I don't think this chapter would be here.

Notes

1   The term "crazy" is often thought to be ableist and many have been arguing that we should be more specific when we use the term. I agree. Because the term often refers to a particularly racialized and gendered notion of defying normative expectations, I have an urge to reclaim it, but it's not really my place, so I'm looking forward to being in solidarity with disability rights scholars and activists in the project of reclaiming the crazy!

2   There are multiple books and articles written about Freeman, a particularly interesting and accessible story is Michael M. Philip's *Wall Street Journal* series (2013).

3   As cited in Brown, p. 447.

4   See my history of chimpanzee research (Gruen, First 100 Chimps, 2006).

5   The week before I gave my presentation at Animaladies in Sydney, Australia (July 2016), two tragically senseless murders of black men by cops had occurred, one in New Orleans, one in Minneapolis, and these were followed by the sniper killing five police officers in Dallas.

6   As reported by Danielle Paquette in the *Washington Post* Wonkblog, July 7, 2016. Available online: https://www.washingtonpost.com/news/wonk/wp/2016/07/07/ the-incredible-calm-of-diamond-lavish-reynolds/?utm_term=.89f0376ddaf6 (accessed February 20, 2018).

7 As reported in the *Daily Mail*, June 30, 2016. http://www.dailymail.co.uk/wires/afp/article-3669253/Nobel-laureate-J-M-Coetzee-speaks-against-animal-cruelty.html
8 As Murdoch (1970) puts it, "Both as act and reason, shopping is public. Will does not bear upon reason, so the 'inner life' is not to be thought of as a moral sphere. Reason deals in neutral descriptions and aims at being the frequently mentioned ideal observer. Value terminally will be the prerogative of the will; but since will is pure choice, pure movement, and not thought or vision, will really requires only action words such as 'good' or 'right'" (page 8).

References

Barad, K. (2012), "Matter Feels, Converses, Suffers, Desires, Yearns and Remembers," in R. Dolphijn and I. Van Der Tuin (eds.), *New Materialism: Interviews and Cartographies*, Ann Arbor, MI: Open Humanities Press.

Bloom, P. (2014), "Against Empathy," *Boston Review*, September 10.

Bloom, P. (2016), *Against Empathy*, New York: Ecco.

Brown, T. H. (1993), "Mental Diseases," in W. F. Bynum and R. Porter (eds.), *Companion Encyclopedia of the History of Medicine* V. 1, London: Routledge.

Dayan, C. (2015), "Feeling into Action," *Boston Review*, September 28.

Dully, C. and Fleming, C. (2008), *My Lobotomy*, New York: Random House.

Freeman, W. and Watts, J. W. (1942), *Psychosurgery: Intelligence, Emotion and Social Behavior Following Prefrontal Lobotomy for Mental Disorders*, Springfield, IL: Charles C. Thomas.

Gluck, J. (2016), *Voracious Science and Vulnerable Animals*, Chicago: University of Chicago Press.

Gruen, L. (2006), First 100 Chimps. Available online: First100Chimps.wesleyan.edu (accessed January 28, 2018).

Gruen, L. (2015), *Entangled Empathy*, New York: Lantern Press.

Murdoch, I. (1970), *The Sovereignty of the Good*, London: Routledge.

NPR (2005), "'My Lobotomy': Howard Dully's Journey" Available online: https://www.npr.org/2005/11/16/5014080/my-lobotomy-howard-dullys-journey (accessed January 28, 2018).

Nuland, S. (2005), "Killing Cures," *New York Review of Books*, August 11, V. 52 #13.

Philip, M. M. (2013), "The Lobotomy Files," *Wall Street Journal*, December 14. Available online: http://projects.wsj.com/lobotomyfiles/ (accessed February 8, 2018).

Rogers, M. (2016), "On Diamond Reynolds after Dallas," July 8. Available online: http://www.publicseminar.org/2016/07/on-diamond-reynolds-after-dallas/ (accessed January 28, 2018).

Wadiwel, D. (2015), *The War against Animals*. Koninklijke: Brill.

Yerkes, R. M. Papers. Manuscripts and Archives, Yale University Library, Group 569, Series II, Box 107.

Young, J. (2015), "Women and Mental Illness," *Psychology Today*, April 22, 2015. Available online: https://www.psychologytoday.com/blog/when-your-adult-child-breaks-your-heart/201504/women-and-mental-illness (accessed February 24, 2018).

Chapter 2

MAKING AND UNMAKING MAMMALIAN BODIES: SCULPTURAL PRACTICE AS TRAUMATIC TESTIMONY

lynn mowson

Prolonged exposure to animal activism and self-enforced witnessing of atrocities committed upon agricultural animals had left me hypersensitive to images of suffering. My research into the dairy industry, and in particular the slaughter of pregnant cows and the treatment of unborn calves in abattoirs, included not only undercover activist descriptions, audio and video footage but also agricultural industry reports and agricultural codes of conduct and regulations—strange dissociative accounts of barbarity.[1] Engaging with this material produced recurrent nightmares filled with the sounds and sights of animal suffering. The constant witnessing of animal death all around me was shattering. Like an undertow, the traumatic knowledge[2] of the lives and deaths of these agricultural animals dragged my sculptural practice into uncharted waters.

At the time I was working on full-scale sculptures of humans, bodies gesturing toward suffering and martyrdom, and I was focused on the particularities of empathic interactions with figurative sculpture.

Enmeshed within my research was an attempt to untangle art historical debates on wholeness and fragmentation of the body, and contemporary art theories that prioritized sculptural body fragmentation and aligned depictions of the whole body with idealism or the saccharine. My sculptural bodies determinedly navigated a visual wholeness and avoided fragmentation with a purpose I would not understand until much later (see Figures 2.1 and 2.2). In retrospect my avoidance of fragmentation and investigation of empathy were driven by my desire to answer other questions about our ethical and exploitative relationships with the nonhuman animal.

Concurrently, my own pregnancy created situations where I needed to justify my decision to raise a vegan child with medical staff and the "concerned"; people become philosophers, nutritionists, and evolutionists around a vegan parent, and accusations of brainwashing my, as-yet unborn, child were not uncommon. Research into vegan nutrition segued into researching the treatment of dairy cows. It was the first time I had become fully cognizant of

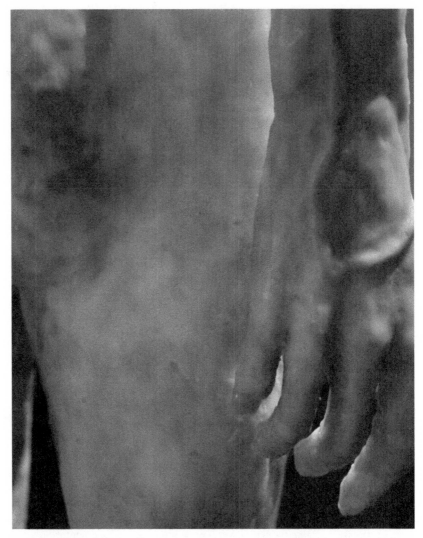

Figure 2.1 lynn mowson, *fleshlump*, 2012–2013 (detail of installation in *beautiful little dead things*, 2014), microcrystalline wax and pigment. Photo: lynn mowson.

the enormous numbers of cows going to slaughter pregnant and there were some horrific statistics; at that time, in the United Kingdom 150,000 pregnant cows were sent to slaughter annually with approximately 40,000 of these in late-stage pregnancy. In a survey in the 1990s, also from the United Kingdom, one slaughterhouse revealed that 23.5 percent of cows were pregnant at the time of slaughter with 26.9 percent of those in their final trimester (Singleton and Dobson 1995). More recent agricultural studies indicate that the numbers of productive and pregnant cows killed worldwide are much higher (Fayemi and Muchenje 2013).[3] Regulations around the slaughter of pregnant cows are applicable only to those cows in the last tenth of their pregnancy (OIE World Organisation for Animal Health 2017).[4] In the last few weeks of fetal development the death these calves experience is mostly dependent on whether by-products are required and gathered from their bodies. For many calves there are no regulations for how they should die. I thought I knew about animal mistreatment, but the sheer numbers caused a visceral response. Beyond the statistics quoted in industry reports there was the personal testimony of witnesses such as Gabriele Meurer,[5] a former official veterinary surgeon in UK abattoirs, who stated:

> What is happening right now in British slaughterhouses is quite simply a scandal. Sometimes when these creatures are hanging on the line bleeding to death, you can see the unborn calves kicking inside their mothers' wombs. I, as a vet, am not supposed to do anything about this. Unborn calves do not exist according to the regulations. I just had to watch, do nothing and keep quiet. It broke my heart. I felt like a criminal. I left the Meat Hygiene Service … completely disillusioned and full of disgust. (Viva! 2014)

My inability to do anything meaningful in response to this knowledge was experienced as an overwhelming trauma, a trauma that returned when I saw people eating cheese or drinking milk around me, building pressure, becoming a silent scream of futility in the face of the enormous normality of animal production and consumption. In the studio, while attempting to avoid the incursions of this knowledge, I simultaneously modified my sculptural figures; they became pregnant and indications of multispecies entanglements such as hair in the wrong places and extra nipples (teats) started to appear (see Figure 2.2).

I laid bare these nonhuman animal clues; the teats were just too perky or were hidden inside out. These were awkward self-conscious attempts to sidestep potentially reductive readings of my work as an iteration of "becoming animal,"[6] or of exhibiting an anthropocentric hybridity that flourished in the artworld and that jarred with my perspective of already "being animal" and my concern with the actual political, ethical, and sociocultural issues at stake when considering agricultural animals.[7]

As a sculptor, the operation of empathy between humans and figurative sculpture has held a prolonged fascination and became increasingly important in my artistic approach. There is a lineage from aesthetic empathy developed in the nineteenth century (Vischer [1874] 1994 and Lipps 1965) through to the phenomenological intersubjective empathy developed in the twentieth century to account for the way in which we understand others to be subjects. The phenomenological empathy articulated by Edith Stein, at the beginning of the twentieth century, is an account of an empathy by which our own somatic and

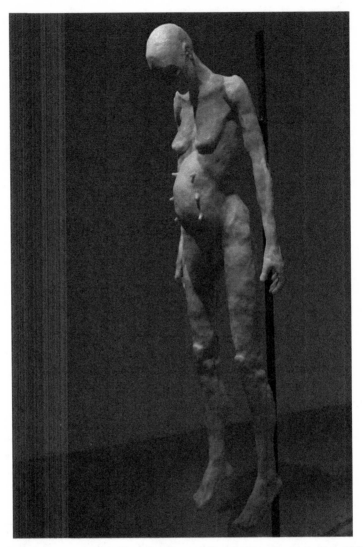

Figure 2.2 lynn mowson, *fleshlump*, 2012–2013 (detail of installation in *beautiful little dead things*, 2014), microcrystalline wax and pigment. Photo: Kerry Leonard.

embodied awareness provides the understanding that there are other bodies in the world; it is how we know that there are other subjects (Stein [1916] 1989). We encounter others through the embodied empathic encounter, an encounter that is a conscious and directed attentiveness to an other. Additionally, it is an ethical encounter that creates an awareness of the absolute alterity of the other, and as such avoids many of the theoretical problems of the models of empathy that depict the process as an unconscious mirroring of the other, a colonizing of the other, or as an unstoppable contagion of the self.[8] This phenomenological empathy is a process: a contingent and precarious encounter, and as such there is always a decision whether to respond and engage with an other (Lipps, 408). Obversely, and just as importantly, a lack of empathy or empathic indifference to others can be an indication of cultural and societal forces that have normalized the withdrawal of empathy toward certain subjects, such as agricultural animals. In this context a withdrawal of empathy is an act of violent erasure of the other.[9]

Lori Gruen's *Entangled Empathy* argues that empathy can have a strong ethical and active role in our relations with animals (2015). Empathy can provide an important pre-linguistic and embodied space of encounter that informs our moral perceptions. As part of the feminist ethics of care tradition, entangled empathy is a process that is based on attending to and responding to animals as subjects with their own lives and needs within our complex economic and cultural systems. My sculptural change in direction was informed by this underlying concern with empathy and an empathic witnessing of the lives and deaths of agricultural animals. By sculpturally blurring the boundaries between the human and nonhuman animal, I tentatively explored and gestured toward our shared mammalian experiences—in particular birth, motherhood, and death. Further, it seemed that I could work toward simultaneously bringing empathic attention to the other while also revealing the violence done to it.

Material changes occurred in my works. I started using latex because it is ephemeral, fragile, and transitory. Latex ages, thins, and eventually decays like our own bodies. I took multiple latex casts from the whole bodies I had made and developed a method to fabricate the latex skins to resemble worn vellum and membranes, dried and treated animal skins devoid of hair. A process of making and unmaking bodies commenced—creating bodies and disassembling them. These adult bodies were subsequently accompanied by a series of *babyforms* and sacs. The *babyforms* were made from the all-in-one baby-suits that my son was outgrowing (see Figure 2.3).

I cast the *babyforms* in multiples, ensuring individuality in each cast, reflecting my desire to bring attention to the subject, the individual within the mass. They became part of the series I eventually entitled *slink* (see Figure 2.4).

"Slink" is a wonderful word meaning lurking, prowling, slithering, skittering, skulking around. Slink leather is a highly valued luxury item, most often used in expensive gloves, religious scrolls, and vellum and it is desirable because it is soft and unblemished. Slink means intrauterine skin; slinks are the skins of animals born prematurely, although calves used for slink leather are not born, but rather left to die in their mother's body or worse.[10] One witness described how an "almost-full-term calf struggled inside and against the mother's body, kicking in desperation, dying a horrible death inside the womb" (Ernst 2013).[11]

The multiple bag-like *sacforms* that complete the *slink* series emerged as a response to the regulations for the slaughter of fetal calves (see Figure 2.5).

Fetal calves can survive their mother's death, and slaughterhouses are advised, if possible, to leave the calves in the uterus for more than five minutes

Figure 2.3 lynn mowson, *slink (babyform)*, 2012, detail, latex, tissue, and string.
Photo: lynn mowson.

Figure 2.4 lynn mowson, *slink*, 2012–2014 (detail of installation in *beautiful little dead things*, 2014), latex, tissue, and string. Photo: Kerry Leonard.

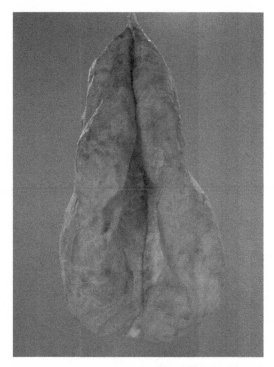

Figure 2.5 lynn mowson, *slink* (*sac*), 2012, latex, tissue, and string. Photo: lynn mowson.

after the mother's throat has been cut: "A foetal heartbeat will usually still be present and foetal movements may occur at this stage, but these are only a cause for concern if the exposed foetus successfully breathes air" (Australian Veterinary Association 2006). These procedures are based on the assumption that the first breath, which oxygenates the brain, produces the ability to perceive pain, and therefore measures should be taken to avoid the calf taking a breath (OIE World Organisation for Animal Health 2017 and van der Valk et al. 2004). When the slaughterhouse collects fetal-bovine serum there becomes a delicate balance of holding the calf in a pre-living stasis while their blood is drained from their beating heart. The Regulatory Code describes how the worker should hold the amniotic sac over the head of a calf to prevent them from taking their first breath:

> Any living fetus removed from the uterus must be prevented from inflating its lungs with air and breathing. This can be done by keeping its head inside the amniotic sac, by clamping its windpipe, or perhaps less satisfactorily, by simulating the amniotic sac by immersing its head in a water-filled plastic bag. These methods would ensure that such fetuses would remain unaware

and therefore would not suffer when exposed to noxious stimuli such as insertion of a 12–16 gauge syringe needle between the fourth and fifth ribs into the fetal heart to allow blood to be collected. (Mellor 2003)

These welfare procedures are not always employed. Timothy Pachirat, in *Every Twelve Seconds*, describes the collection of fetal-bovine serum:

> Sometimes out of the pipe in the wall an oblong gray mass shoots that is not a lung, kidney, windpipe or liver … the white-helmeted worker walks over, picks up the object … cuts into the grey mass. There will be a fetus inside, with smooth, slick skin, and clearly marked hide patterns. Raising the fetus up by the neck and hind legs, the man swivels … and pushes the fetus's mouth onto one of the protruding hooks … he uses two hands to stick another hook into the fetus's anus. The fetus now hangs suspended by its mouth and anus, and the worker makes an incision in the neck area, bringing a bottle with a straw … to the incision. (2011, 79–80)

Calves are kept on the brink of life and death in order to collect blood, a balance between animal welfare and production necessity. Research has indicated that fetal calves are likely to be exposed to discomfort and pain; studies on fetal resistance to suffocation reveal that pain can be experienced until brain damage occurs due to lack of oxygen. In fact bovine fetuses are considered able to feel pain from about a third of the way into their gestation, when the neural tube develops into a functioning brain, and some studies suggest that due to the way the brain is developing they might actually feel pain more extremely (Jochems et al. 2002). My latex *sacforms* evoke the meaty nourishing placenta, the emptied caul, and the amniotic sac put to the most brutal use in the slaughterhouse. Fetal calves exist in a liminal space, between life and death, between waste by-products and useful resource, between disregard and regulation, creatures whose first breath must be smothered while they are kept alive and their blood is drained and later their skins are collected.

Through the witnessing of this traumatic knowledge, the work *slink* emerged. It was not consciously directed, and often resisted, propelled from an unconscious material working through and regurgitation of my garbled and somewhat hysterical traumatic thoughts: a mass of assimilated and unassimilated, compressed and compounded scraps of images both seen and imagined; the fetal calf kicking as he/she was cut from his/her bled-out mother, her body hanging upside down, and the cold emotionless descriptions of horrific procedures and processes that reemerged as detailed nightmares and flashbacks.

The changes that occurred through this traumatic working practice often surprised me, for example the violence I enacted on my sculptures proved to be something of a shock. Removing the latex from the mold was a rough and violent task that involved cutting, tearing, and pulling the skins from the little baby forms. To me this process of violent skinning echoed the disassembling of animal subject that occurs both literally and linguistically, and I found it a challenge to undertake, as my studio notes attest:

> Skinning days. The bodies are taken down from their hanging position, and rubbed with talc, rubbing into the crevices, the pits, the groin. A sharp scalpel makes a cut through the skin, and then the process of skinning commences, pulling and rubbing down, pulling, stretching, releasing, it is an ugly task, on par with creating meat lumps from the creatures. I'm sometimes callous with my little objects. Sometimes their grip on me is too strong, their tenderness, fragility and exposure is too painful to bear. I become meat worker, butcher, in order to do this work, and I make amends [regrettably useless] with small objects. I counterbalance the butchering with the endless work of reparation. (Studio notes)[12]

Concurrent with the development of *slink* I had started to violently dismember my whole wax bodies and animal portraits (*creatures*), butchering them into fleshy meaty lumps. While many of these *fleshlumps* were pushed beyond recognition, others bear traces of their origin—a nostril here, a teat there (see Figure 2.6).

I questioned the impact of this unmaking of the body; was it an act of violent erasure, was it a reduction of a subject to an object? Had I deliberately excluded the possibility of an empathic encounter with these objects or did

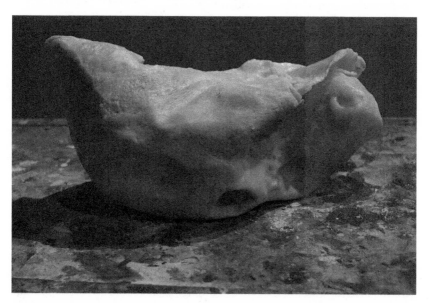

Figure 2.6 lynn mowson, *fleshlump*, 2011–2013 (detail of installation in *beautiful little dead things*, 2014), microcrystalline wax and pigment. Photo: Kerry Leonard.

the remaining traces of the body create empathic opportunities? I followed the violence of skinning *slink* and dismembering the *fleshlumps* with a phase of care and reparation: either carefully resurfacing the waxes or gently washing and then sewing the latex forms together. Sewing was a small intimate practice of care, a futile and impossible reparation task, as if attempting to atone for brutality in my work and the world beyond. The disassembling of my whole sculptural bodies from presence into absence led to a radical reevaluation of my processes and practice, and a consideration of the affect of trauma on my works.[13]

Trauma affecting animals' advocates is complex and has only in recent years gained attention as an issue for animal activism.[14] As an activist it was, and in some cases still is, inadvisable to mention one's own trauma in relation to vastness of animal suffering, and one feels compelled to continue to endlessly bear witness to acts of cruelty, to avoid looking away. Rational argument not emotion is lauded, although these rational arguments have not stopped the inexorable expansion and industrialization of animal agriculture. Nor have they conquered the seemingly subjectively embodied collaborators in cruelty, taste and pleasure.

> Sometimes I stand at the edge of butchers' shops, normally I rush by, their smell envelops me, covers me in a layer of flesh and fatty tissue, the thick smell of cold death. I wonder at how normal this all seems to people, these lumps of bodies laid out, creatures that had lives given and taken from them for a meal. How invisible yet visible the atrocity is. This atrocity that is invited into our homes and our mouths. (Studio notes)

As late as the 1990s, trauma was considered to be an event outside the range of normal human experience and most commonly used to describe the symptoms of Holocaust or war survivors. The definition of trauma was amended by the American Psychiatric Association in 1994 to become any event "involving 'actual or threatened death or serious injury, or a threat to the physical integrity of self or others'" (Bryant 2006, 94). Animal activists experience a threat to the physical integrity of an other when they witness the suffering of animals through audio-visual or other methods (95–96). Activist and writer pattrice jones states: "even the stringent diagnostic criteria for PTSD acknowledge that witnessing death, injury, or the threat thereof to another can be a traumatic experience. In fact, some researchers have found that witnessing—and being helpless to stop or prevent—harm to another can have a greater traumatic effect than being a victim of violence oneself" (2007, 93).

Traumatic affects include experiences of anguish, guilt associated with not having prevented harm or death, and "sadness, grief, depression, anxiety, dread, horror, fear, rage, and shame; intrusive imagery in nightmares, flash backs, and images; numbing and avoidance phenomena; cognitive shifts in viewing the world and oneself, such as suspiciousness, cynicism, and poor self-esteem"

(Valent 2002, 19). Those affected can have "intrusive thoughts about the violence; hypersensitivity to triggers of thoughts and memories of the violence; volatile personal relationships; persistent and involuntary emotional numbing or unregulated emotional extremes of outbursts and low affect; severe mistrust of others" (Bryant, 102). Trauma can also cause hypermnesia, which is a strong recall or vivid recall of memories, as well as those memories being "garbled" because of the difficulty of processing traumatic memories into everyday life.

jones coined the term "aftershock" to describe the aftereffect of a traumatic event and observed post-traumatic stress disorder in animal advocates as basically a normal reaction to extraordinary experiences (75–76). *Aftershock* focuses on recovery; however, jones notes that while general trauma therapy focuses on coming to terms with and integrating trauma into one's worldview, this is almost impossible for the animals' advocate. It is the 'reality' of everyday life, jones contends, that continues to inflict trauma on the animal advocate; "everyday life can be similarly nightmarish for those who have undone the socialization that leads us to see cadavers as 'meat'" because "vegans, unlike flesh-eaters, never stop noticing the violence inherent in meat" (90, 149). We are surrounded in day-to-day life by the constant reminders of what I call the ag-trocities (atrocities caused by the animal agricultural industry) against animals. For the affected witness, meat, dairy, and leather products are no longer food and clothing products: they are flesh products, bodies and parts of once living beings. Animals are co-opted in advertising into promoting their own fleshy bodies to be consumed.[15] Supermarket shelves are filled with parts of animal bodies, and friends and family eat these foods in front of you: "vegans are constant witnesses to the horrors of carnism, [and] must live in a world that daily offends their deepest sensibilities" (Joy 2012).

Advertisements and children's books and toys depict unrealistic pictures of farms: happy cows in the fields, pigs smiling, and chickens roaming free. In children's books animals, particularly agricultural animals, start as empathic fellow creatures and playmates, and then are gradually moved away from us, moved onto the farm and turned into product machines. Raising a vegan child meant that I was hyperaware of the normalizing narrative processes of carnism, and the gradual disassociation of empathy articulated so well in Jane Legge's *Learning to Be a Dutiful Carnivore* (1969):

Eat the flesh from "filthy hogs"
But never be unkind to dogs.
Grow up into double-think—
...
They only come on earth to die,
So eat your meat, and don't ask why.

Fiona Probyn-Rapsey recently articulated the term "animalady" to "gesture at a state of profound disease in the face of destructive human–animal relationships but with the view that such disorders can provoke positive transformations" (2014, 16). In this concept of "animalady" both trauma and the transformative possibilities of traumatic affects can coexist; in this context my trauma resulted in the production of the new works: *slink* and *fleshlumps*. Both of these series are a transformation of my traumatic knowledge into what I consider to be a form of sculptural testimony. Testimony transforms traumatic knowledge into words/sound/text/object. It is essentially outwardly directed toward another, offered to another, and ideally reaches a responsive audience with a shared communicative language: visual, verbal, or textual.[16] However, my sculptural testimony is often experienced across the experiential abyss between the vegan and the carnist, across a seemingly impassable divide.

As traumatic testimony, my material working through was overstretched to encompass the traumatic memories of the abuses in the slaughterhouses and factory farms; it was further burdened with the weight of the whole sociocultural processes by which these events are normalized—carnism. In facing the overwhelming sociocultural mechanisms of carnism my work became somewhat contradictory and hysterical as I fought also to suppress the incursion of agricultural animal lives and deaths into my work. In trying to discuss the underlying tensions in my work I found myself positioned as the "radical vegan freak," the "vegan killjoy" (Twine 2014),[17] the "out-there irrational-nutty-screw-loose-mushy-softy-weirdo-bunny-hugger-animal-lover." Moreover, the reception of my sculptural testimony was never clear-cut, sometimes animal advocates rejected my work: the message is not clear enough, the works are too meaty, they look too much like real skin. Similarly, I too questioned the aesthetics of my representations of animal skin and flesh, my representation of skinned bodies and fleshy lumps. To continue this traumatic making and unmaking of bodies, this sculptural testimony, I needed to surrender to the possibilities of communicative mistakes, failures, misreadings, and misunderstandings.[18]

While the emergence of traumatic knowledge in my practice had been initially undesirable, it focused my attention on how to align my animal advocacy with my art, since it had become very clear that one would not leave the other alone. So while *slink* and *fleshlumps* emerged somewhat unbidden from trauma, with my series *boobscape* (2016–2017) I determined from the outset to utilize the productive force of "animalady." My focus on dairy cows continued, but moved from the slaughterhouse to the conditions of their lives, and the exploitation of their motherhood for the production and consumption of milk. For me there is no surprise that the word "milk" has been used to describe exploitation for profit for centuries; to milk is to tap, exploit, bleed, drain, extort, extract, suck dry, wring, elicit, empty, exhaust, fleece, press, pump, siphon, use, draw off, impose on, let out, take advantage of, take out.

The horrors of the dairy industry are so extensive that here I outline only a few of the elements that fomented *boobscape* (see Figures 2.7 to 2.10).

Figure 2.7 lynn mowson, *boobscape*, 2016–2017 (image used for Animaladies postcard collection, 2016), latex, tissue, and string. Photo: lynn mowson.

Traditional dairy farms are decreasing; dairying is big business, but only for those who farm intensively. Some mega-dairies can house over 15,000 cows. Cows are kept indoors with their feed manipulated and zero-grazing opportunities causing lameness, hoof problems, teat tramp, and mastitis. Intensive farming leaks into the environments millions of liters of slurry need to be disposed of, often poorly, causing water and air pollution and wildlife diseases.[19]

While the average life span of a wild cow is around twenty to thirty years, commercial dairy cows go to slaughter around their fifth to eighth year, worn out. Milk making is physically hard work and dairy cows are expected to produce milk for up to ten months of the year. This depletes the body of minerals and nutrients, meaning contemporary dairy cows are in a constant state of metabolic hunger. They get a break from milk production only during the last fifty to sixty days of pregnancy—the "drying off" period to allow the udder to recover and for the treatment of mammary infections.

Selective breeding has increased production of milk; a calf would normally drink around eight times daily, keeping the production of milk stable, and the udders and teats in good condition. Mastitis is one of the leading causes of antibiotic use in dairy farming and for the culling of pregnant cows (Food and Agricultural Organization of the United Nations, 1989). Mastitis is an inflammation of the mammary gland, with a high economic cost due to decreased milk quantity and quality, and fever and depression in cows (Jones 2009). Milking machines transmit the mastitis pathogens to other members of the herds, and they also frequently damage teats. And as cows age their propensity for mastitis increases. Controls for mastitis include tail docking, injections of antibiotics into the teats in the dry phase, and the burning of udder hair—known as flame clipping. Flame clipping is increasingly replacing other more labor-intensive forms of hair removal, as hairy udders are thought to gather more dirt and require more cleaning, and they impair robot efficiency, but error rates can be high, and farmers report "flare ups" requiring dousing (Gamroth et al. 2000). Mastitis not only affects the udder internally but also causes ulcerations on the flesh, and teat sores and cracks.

The repetitive exhausting pregnancies, the limping lameness, the extended, overburdened ulcerated diseased udders, the memory of the cow's foot stamping as the flame goes over their udders and lingers a little too long, and the endless removal of calves were the starting point for *boobscape*. The exploitation of motherhood in the dairy industry emerged in the studio as multiple mammary

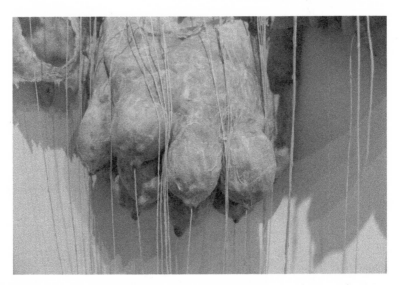

Figure 2.8 lynn mowson, *boobscape*, 2016–2017, latex, tissue, and string. Photo: Mary C Holmes.

glands of humans, bovines, and other fellow-creatures, these multispecies mammaries indicating the ongoing embodied and empathic approach of my work. Single udders became multibreasted and nipples elongated into teats. The breast skin is dry and aged, weathered and sore, veins bulging and ulcerated skin peeling, there are sore breasts, dried out empty breasts, worn nipples, leaky lactating breasts, breasts blurring between species, multiple breasts indicating the endless cows and the endless reproduction in the dairy industry. The surrounding skin is papery thin and soft like membranes and dried fleshless skins, like unprocessed slink skins as a material testimony for those little lives lost in the dairy industry (see Figure 2.9).

boobscape makes possible embodied and empathic connections by obscuring the boundaries and the spaces between the mammary glands shared between mammalian species, increasing the potentiality for a viewer to experience an embodied empathy mediated through sculpture. I hope that one might, on encountering these mammary forms, clasp one's hand to one's breast, and think "ouch." That one might imagine the swollen bruised hot breasts, veins bulging,

Figure 2.9 lynn mowson, *boobscape*, 2016–2017, latex, tissue, and string. Photo: Kerry Leonard.

the awful searing pain of mastitis, the feel of nipples raw to touch, skin flaking. That one might empathically respond to the materiality and form, and in doing so recognize those things we share with our fellow creatures,[20] our entangled embodiment and potentially shared physical memories of childbirth and breast-feeding. No longer do I tentatively place teats on human forms, now I muddy the boundaries between species, when teats become nipples and breasts become udders, I evoke our entangled nurturing and exploitative relations with other creatures and I nod to visual lineage of humans suckling animals and animals suckling humans. I witness a history that has human–animal suckling closely entwined, although these are no utopian spaces, human needs surmount animal needs.

boobscape alludes to possible monstrous milking machines, animals reduced to mammaries, the multiplicity of the herd, multibreasted goddesses, boundary crossings and infringements between human and nonhuman animals, to the potent abject and transgressive possibilities of milk and motherhood. There are breasts that bud and form and mutate; there are supernumerary teats, prepubescent nipples.

Calves can be born with an extra teat, a supernumerary teat, a sprig or a web teat; these are removed because they can interfere with future milk collection practices.[21] They can be cut off with scissors, the codes recommend but do not enforce pain relief.[22] Teats are unwanted elements in leather production, the waste by-products of the animal body. Thus my process of sculptural testimony continues to strive to include all this traumatic knowledge and more.

Figure 2.10 lynn mowson, *boobscape*, 2016–2017, latex, tissue, and string. Photo: Kerry Leonard.

boobscape lingers discordantly between beauty and horror, attempting a balance between empathy and erasure through making and unmaking mammalian bodies (Figure 2.10). While I hope my sculptural offering opens the possibility of an empathic encounter, my struggle with the misunderstandings and misreadings of my sculptural testimonial continues. As often as not, I find myself standing at the edge of that abyss of incomprehension—that yawning gap between the ubiquity of a glass of milk and my anger and bewilderment at the horrors of the dairy industry, mumbling: "That is not it at all, 'That is not what I meant, at all" (Elliot 1963, 17).

Acknowledgment

Special thanks go to creative writers and collaborators Hayley Singer and Sue Pyke, whose ruminations in my studio were influential on the development of boobscape. I would like to thank Melissa Boyde, Maddi Boyd and Yvette Watt for including me in the Animaladies exhibition. Fiona Probyn-Rapsey for her ongoing enthusiasm and encouragement of artists working in this field. Curators Kathryn Eddy, Janell O'Rourke and L.A. Watson for including slink and boobscape in the exhibition SPOM: The Sexual Politics of Meat at The Animal Museum, LA, and of course Carol J. Adams whose work inspired the exhibition.

Notes

1　Two influential studies subsequently consolidated information on the lives of dairy cows (Butler 2014, Gellatley 2014). See also *The Life of the Dairy Cow: A Report on the Australian Dairy Industry* (Voiceless 2015).
2　"The death of nonhumans not only violates specific nonhuman animals, it is a violation of the humane desire for the good and the just. The death of nonhumans is always on hand for animal rights activists. *This is traumatic knowledge*" (Adams 2004, my emphasis).
3　These studies are produced because of the perceived economic waste of late-stage calves.
4　The Code advises against transportation and slaughter for late-stage pregnancy. However, in the event that late-stage slaughter occurs the following rules apply: "Foetuses should not be removed from the uterus sooner than 5 minutes after the maternal neck or chest cut, to ensure absence of consciousness" (OIE World Organization for Animal Health 2017). Otherwise they advise leaving the fetus in the body until dead.
5　Gabriel Meurer's (2007) undercover video footage of the slaughter of pregnant cows is available on YouTube.
6　The concept of "becoming animal" had a significant impact in the visual arts (see Thompson 2005). As an example of the anthropocentrism I find problematic, I

quote the overview for the book, which states: "In an age when scientists say they can no longer specify the exact difference between human and animal, living and dead, many contemporary artists have chosen to use animals in their work—as the ultimate 'other,' *as metaphor, as reflection*" (The MIT Press 2018, my emphasis).

7   These important conversions around the interpretations of "becoming animal" and the post-human are out of the scope of this chapter; for an example of the issues raised, see Iveson (2013).

8   The theoretical debates around the term "empathy" are complex; however, it is important that Stein's intersubjective empathy be distinguished from the instinctual or automatic mirroring as considered by proponents of mirror-neuron empathy (see Rizzolatti et al. 1992 and Gallese et al. 1996).

9   See Dean for a discussion of empathic indifference as an act of cruelty and a "violent erasure" (2004, 104).

10 Slink is also made from lambskin sourced from the bodies of newly born and deceased lambs, although this is less desirable for vellum due to size.

11 Stephanie Ernst provides this textual account from Meurer's video of the slaughter of a pregnant cow (2013).

12 I write personal notes while working in my studio. Short extracts are included in this chapter.

13 For a discussion of these works in terms of absence, see mowson (2015).

14 This is indicated by the emergence of new terms to describe the trauma affecting animal advocates. Psychologist Melanie Joy uses the term "carnism-induced trauma": "it is our assertion that carnism, like other atrocities, causes mass traumatization of not only the primary victims (farmed animals) but of secondary victims (humans) as well. For clarity and accuracy, we describe the traumatization resulting from carnism as carnism-induced trauma" (2012). Recently, the term "Vystopia" has been coined by psychologist Clare Mann: "1. Existential crisis experienced by vegans, arising out of an awareness of the trance-like collusion with a dystopian world. 2. Awareness of the greed, ubiquitous animal exploitation, and speciesism in a modern dystopia" (2017).

15 Ben Grossblatt refers to "any depiction of animals that act as though they wish to be consumed" as "suicide food" (2011).

16 The issue of trauma and language and the failures of language to communicate that are amplified by trauma are beyond the scope of this chapter. For accounts on testimony, see Derrida (2000), Caruth (1995), LaCapra (2001), and Trezise (2013).

17 Special thanks go to Richard Twine for articulating the political agency and positive disruptive role of the vegan killjoy (2014).

18 For example, my series *fleshlumps* created hunger for one carnist friend and revulsion in a vegan colleague, neither of which responses were my intention.

19 This has been widely reported and monitored by government agencies and industry bodies (see United States Environmental Protection Agency 2001; Dairy Australia 2008; Dairy Industry Profile Agriculture Victoria 2014; and other state-based departments responsible for environment and primary industries in Australia).

20 I use this term in line with Cora Diamond's concept of nonhuman animals as fellow creatures. Animals, Diamond contends, are our "fellows in mortality"; we share fear and pain; birth and death; social and emotional bonds with animals (Diamond 1978).

21 "While calf rearing practices are basically the same for conventional or robotic milking, removal of extra teats and disbudding are more important for calves that will be milked automatically because extra teats can slow down robotic cup attachment" (Kerrisk 2015).

22 The RSPCA recommends this procedure is done prior to three months old and with anesthetic (2009). In the United Kingdom, a calf is protected once they reach three months old (Department for Environment, Food and Rural Affairs, 2003).

References

Adams, C. J. (2004), "Home Demos and Traumatic Knowledge," *SATYA*. Available online: http://www.satyamag.com/mar04/adams.html (accessed August 1, 2014).

Agriculture Victoria (2014), "Dairy Industry Profile," Available online: http://agriculture.vic.gov.au/agriculture/dairy (accessed February 1, 2017).

Australian Veterinary Association (2006), *Policy 85: Fetal Bovine Serum Collection.* Available online: http://www.ava.com.au/policy/85-fetal-bovine-serum-collection (accessed June 1, 2012).

Bryant, T. L. (2006), "Trauma, Law, and Advocacy for Animals," *Journal of Animal Law and Ethics*, 1 (63): 63–138.

Butler, J. (2014), *White Lies, A Viva! Health Report*, J. Gellatley (ed.), Viva! Health: Bristol.

Caruth, C. (ed.) (1995), *Trauma: Explorations in Memory*, Baltimore and London: The Johns Hopkins University Press.

Dairy Australia (2008), "Effluent and Manure Management Database for the Australian Dairy Industry." Dairy Australia. Available online: http://www.dairyaustralia.com.au/Environment-and-resources/Soils-nutrients-and-effluent.aspx (accessed February 1, 2017).

Dead Unborn Calves (2007) [Video]. Dir. G. Meurer. Available online: https://www.youtube.com/watch?v=Zytvvi4Q-aw (accessed January 1, 2012).

Dean, C. J. (2004), *The Fragility of Empathy after the Holocaust*, Ithaca; London: Cornell University Press.

Department for Environment, Food and Rural Affairs (2003), *Code of Recommendations for the Welfare of Livestock: Cattle*, PB7949. United Kingdom.

Derrida, J. (2000), " 'A Self-Unsealing Poetic Text': Poetics and Politics of Witnessing," in M. P. Clark (ed.), R. Bowlby (trans.), *Revenge of the Aesthetic: The Place of Literature in Theory Today*, 180–207, Berkeley: University of California Press.

Diamond, C. (1978), "Eating Meat and Eating People," *Philosophy* 53 (206): 465–479.

Elliot, T. S. (1963), "The Love Song of J. Alfred Prufrock," *Collected Poems 1909–1962*, London: Faber and Faber.

Ernst, S. (2013), "Pregnancy at Slaughter: What Happens to the Calves?" Available online: http://www.all-creatures.org/articles/ar-pregnancy.html (accessed June 1, 2013).

Fayemi, P. O. and Muchenje, V. (2013), "Maternal Slaughter at Abattoirs: History, Causes, Cases and the Meat Industry," *SpringerPlus* 2 (125): 1–7.

Gallese, V., Fadiga, L., Fogassi, L., and Rizzolatti, G. (1996), "Action Recognition in the Premotor Cortex," *Brain* 119 (2): 593–609.

Gamroth, M., Downing, T., and Peters Ruddell, A. (2000), *Flame-clipping Udders on Dairy Cows*, EM 8755, Oregon State University Extension Service.

Gellatley, J. (ed.) (2014), *The Dark Side of Dairy, A Viva! Report*, Viva! Health: Bristol.

Grossblatt, B. (2011), "Five Years: An Announcement," *suicidefood*. Available online: suicidefood.blogspot.com (accessed June 1, 2016).

Gruen, L. (2015), *Entangled Empathy*, New York: Lantern Books.

Iveson, R. (2013), "Deeply Ecological Deleuze and Guattari: Humanism's Becoming-Animal," *Humanimalia: A Journal of Human/Animal Interface Studies* 4 (2): 20–40.

Jochems, C. E., Van Der Valk, J. B., Stafleu, F. R. and Baumans, V. (2002), "The Use of Fetal Bovine Serum: Ethical or Scientific Problem?" *Alternatives to Laboratory Animals* 30: 219–227.

Jones, G. M. (2009), *Guidelines to Culling Cows with Mastitis*, Virginia Cooperative Extension Publications, Publication no. 404-204: 1–3. Available online: http://pubs.ext.vt.edu/404/404-204/404-204.html (accessed January 1, 2016).

Jones, P. (2007), *Aftershock: Confronting Trauma in a Violent World, a Guide for Activists and Their Allies*, New York: Lantern Books.

Joy, M. (2012), "Carnism-Induced Trauma: How to Avoid Burning Out as an Animal Rights Activist," republished in All-Creatures. Available online: http://www.all-creatures.org/articles/act-ab-carnism-induced-trauma.html (accessed June 1, 2016).

Kerrisk, K. (2015), "Raising Cows for AMS," *Future Dairy*. Available online: http://futuredairy.com.au/raising-cows-for-ams/ (accessed June 1, 2016).

LaCapra, D. (2001), *Writing History, Writing Trauma*, Baltimore: Johns Hopkins University Press.

Legge, J. (1969), "Learning to Be a Dutiful Carnivore." Available online: http://www.all-creatures.org/poetry/ar-learning.html (accessed June 1, 2014).

Lipps, T. (1965), "Empathy and Aesthetic Pleasure," in K. Aschenbrenner (ed.), *Aesthetic Theories: Studies in the Philosophy of Art*, 403–412, Englewood Cliffs: Prentice-Hall.

Mann, C. (2017), "Vystopia." Available online: http://vystopia.com/index.html (accessed December 1, 2017).

Mellor, D. J. (2003), "Guidelines for the Humane Slaughter of the Fetuses of Pregnant Ruminants," *Surveillance* 30 (3): 26–28.

Mowson, l. (2015), "beautiful little dead things," in K. Eddy, L. A. Watson, and J. O'Rourke (eds.), *The Art of the Animal: Fourteen Women Artists Explore the Sexual Politics of Meat*, 107–116, Brooklyn: Lantern Books.

OIE World Organisation for Animal Health (2017), *Terrestrial Animal Health Code*. Available online: http://www.oie.int/international-standard-setting/terrestrial-code/access-online/ (accessed June 1, 2017).

Pachirat, T. (2011), *Every Twelve Seconds: Industrialised Slaughter and the Politics of Sight*, New Haven, CT: Yale University Press.

Probyn-Rapsey, F. (2014), Review Article: "Multispecies Mourning: Thom van Dooren's Flight Ways," *Animal Studies Journal* 3 (2): 4–16. Available online: http://ro.uow.edu.au/asj/vol3/iss2/3 (accessed June 1, 2016).

Rizzolatti, G., Di Pellegrino, G., Fadiga, L., Fogassi, L., and Gallese, V. (1992), "Understanding Motor Events: A Neurophysical Study," *Experimental Brain Research* 91: 176–180.

RSPCA (2009), "Invasive Farm Animal Husbandry Procedures," *Position Paper B4*. Available online: http://kb.rspca.org.au/files/2/ (accessed June 1, 2016).

Singleton, G. H., and Dobson, H. (1995), "A Survey of the Reasons for Culling Pregnant Cows," *The Veterinary Record* 136 (7): 162–165.

Stein, E. ([1916] 1989), *On the Problem of Empathy*, W. Stein (trans.), Washington: ICS Publications.

The MIT Press (2018), "Becoming Animal: Overview." Available online: https://mitpress.mit.edu/books/becoming-animal (accessed June 1, 2016).

Thompson, N. (ed.) (2005), *Becoming Animal: Contemporary Art in the Animal Kingdom*, Cambridge; London: MIT Press.

Trezise, T. (2013), *Witnessing Witnessing: On the Reception of Holocaust Survivor Testimony*, New York: Fordham University Press.

Twine, R. (2014), "Vegan Killjoys at the Table—Contesting Happiness and Negotiating Relationships with Food Practices," *Societies* 4 (4): 623–639.

United States Environmental Protection Agency (2001), "Notes from Underground," *EPA Region 9 Water and Underground Pollution Newsletter*. Available online: https://nepis.epa.gov/. Department of Primary Industries (accessed June 1, 2016).

Valent, P. (2002), "Diagnosis and Treatment of Helper Stresses, Traumas, and Illnesses," in C. R. Figley (ed.), *Treating Compassion Fatigue*, 1–17, New York: Taylor and Francis.

Van Der Valk, J., Mellor, D., Brands, R., Fischer, R., Gruber, F., Gstraunthaler, G., Hellebrekers, L., Hyllner, J., Jonker, F. H., Prieto, P., Thalen, M., and Baumans, V. (2004), "The Humane Collection of Fetal Bovine Serum and Possibilities for Serum-Free Cell and Tissue Culture," Workshop Report *Toxicology in Vitro* 18, 1–12.

Vischer, R. ([1874] 1994), "On the Optical Sense of Form: A Contribution to Aesthetics 1873," in H. F. Mallgrave and E. Ikonomou (eds.), *Empathy, Form, and Space: Problems in German Aesthetics, 1873–1893*, 89–124 Texts & Documents. Santa Monica: University of Chicago Press.

Viva! (2014), "Throwaway Lives: Stop the Massacre of Pregnant Animals," *Briefing Notes*. Available online: http://www.viva.org.uk/what-we-do/pregnant-cow-massacre/briefing-notes (accessed 2014).

Voiceless (2015), *The Life of the Dairy Cow: A Report on the Australian Dairy Industry*, Voiceless, The Animal Protection Institute.

Chapter 3

THERE'S SOMETHING ABOUT THE BLOOD ... : TACTICS OF EVASION WITHIN NARRATIVES OF VIOLENCE

Nekeisha Alayna Alexis

Ours in such circumstances, are kindly cared for, and are never considered a burden; our laws are, generally speaking, humane and faithfully administered. We have enactments which not only protect their lives, but which compel their owners to be moderate in working them, and to ensure them proper care as regards their food.

—Eastman (1852, 70–71)

They were happy chickens. They had the run of the yard, and they took advantage of it. They grazed on the lawn and lolled about in the breeze under the treehouse. When we worked in the garden, our chickens were right beside us, companions raking the earth. We treated those birds right ...

—Roth (2016, 25)

Narratives of violence about conscientiously consuming ethically raised farmed animals are perplexing. On one hand, those who advocate humane farming and do-it-yourself slaughter testify to a host of positive outcomes such as cultivating reverential eating habits, increasing compassion toward other animals, and restoring human relationships in a broken agricultural system. Yet proponents of knowing and killing your dinner as "a strong corrective to dislocation and alienation in our industrial food system" (Kaminer 2010, MB1) often leave unexamined the substantial contradictions inherent within this action. The unwillingness to grapple with the tensions in this ritual are especially curious since the intent of slitting a sow's throat or dismembering a hen's body is to unflinchingly face the bloodshed that converts an individual into food. The oversight is also strange since this form of conscious omnivorism has gained visibility amid and in response to increased mainstream and liberationist vegan agitation against using other animals as food.[1]

Interestingly, this practice of redirecting questions of power and violence toward other tangential considerations is not unique to do-it-yourself slaughter writings. On the contrary, the effort to reframe the grotesque as benign is also present in at least one other narrative of violence: that of plantation romances. In these nineteenth-century works of fiction, authors used various "polemical strategies" to reassert the legitimacy of slavery in the face of abolitionist pressures (Railton n.d.[a]). Although plantation romances and conscious omnivore narratives are dissimilar in significant ways, these seemingly disparate genres rely on common storytelling devices to make their case. In this chapter, I identify their overlapping tactics of evasion and interrogate how they work to make palatable again destabilized systems of domination.

To begin, I review a handful of do-it-yourself slaughter reflections and humane farming accounts with a focus on short articles published for popular audiences. Here, I describe some of the techniques that divert attention from the discrepancies within humane agriculture and agribusiness. I then outline how these tactics function in two plantation romances, specifically the anti-Tom novels *Aunt Phyllis's Cabin; or, Southern Life As It Is* (1852) and *The Planter's Northern Bride* (1854). In examining these texts together, I make evident some of the ill-logics dominant groups employ to justify subjugating other persons. By extension, I challenge the idea of conscious omnivorism as an antidote to exploitative relationships with other animals and highlight the insidious nature of its arguments.

Before proceeding, I want to clarify my interest is not in comparing the conditions facing humanely farmed animals in the present with those slaves endured throughout the sixteenth and seventeenth centuries in the United States. Nor am I suggesting that these, or other conscious omnivores, are drawing on inspiration from plantation romances or advocating for the enslavement of black people. Instead, I am fascinated by the mechanics of simultaneously writing *and not writing about* power within acts of interpersonal and systemic violence, especially when the validity of those behaviors is upset. Reading humane farming accounts and do-it-yourself slaughter reflections alongside plantation romances is one way I seek to untangle and understand this dynamic.

The narrative terrain: Liberationist veganism and conscious omnivorism

In recent years, a narrative contest has emerged between vegans and conscious omnivores as both movements respond to the crises caused by industrial farming. In general, vegans criticize factory farming for its deplorable treatment of other animals and its devastating effects on the environment and human health, and, at minimum, choose a plant-based diet to lessen those conditions. However, a particular wing I refer to as liberationist vegans also see these calamities as resulting from the prevailing views that humans are superior to and separate from other animals; that other animals are destined or designed for human use; and that the legal, economic, political, and social constructs emerge from and perpetuate these perspectives. By refusing to eat other animals, advocating for plant-based diets, and working to end factory farming among other activism, liberationist vegans resist a larger paradigm that reduces other animals to human instruments, denies them equitable ethical consideration, and provides license to kill them. For liberationists, an overarching goal is dismantling the confused categories of "human" and "animal" such that other animals can, as much as possible, be self-determining individuals in and for their communities, and people's use and abuse of other creatures is no longer normalized.

As mainstream and liberationist vegans have exposed gratuitous violence against animals and workers, and ecological irresponsibility across major flesh-food industries, they have also troubled the landscape around animal agriculture and agribusiness. Undercover investigations have led to slaughterhouse closings and legislation against some of the most egregious farming techniques. Meanwhile, vegan campaigns have contributed to a rise in plant-based alternatives and eating (Strom 2017), reduced milk sales resulting in billions of lost dollars (Yu+ 2017), and a decade-long dip in beef and other meat consumption. At the time of writing, 6 percent of US consumers identify as vegan—a 5 percent gain in that foodway in only three years (GlobalData 2017). Additionally, leading meat producers like Tyson Foods and Maple Leaf Foods have begun investing in plant-based companies; ninety-year-old Elmhurst Dairy has switched to an all plant-based production, and there are more reports of small-scale animal farms converting to vegetable operations and sanctuaries.

Although meat production in the United States is bouncing back (Sawyer 2016), the case against flesh-foods remains strong. The World Health Organization's report on the potential carcinogenicity of processed meats (Bouvard et al. 2015) and UN reports that reduced meat consumption can stave off climate change (IPCC 2015) have challenged the industry. Documentaries like *Cowspiracy* and *What the Health?* and the film *Okja* are also popularizing plant-based eating. Although the dent made by animal advocates is tiny compared to the billions of cows, sows, hens, and others still exploited and killed annually in the United States alone, debates about the validity and viability of animal agriculture are more commonplace. It is in this context of actual and perceived upset, when "the social norm of meat-eating seems to be losing its rigidity" (Gutjahr 2013, 380), that humane farming and do-it-yourself slaughter have gained traction.

Unlike most consumers, conscious omnivores see farmed animals as unique individuals with distinct personalities, desires, and needs, and condemn standard flesh-food production for its abuses. They also "question the normality of the consumption, or high consumption of meat" and criticize other meat, egg, and dairy eaters for ignoring the consequences of current food practices (Gutjahr 2013, 381). Essential to their approach is acknowledging that living, thinking, feeling animals die to accommodate their diet and moderating their intake in light of that awareness. Adherents insist on compassionate treatment of animals from birth through death, including appropriate living conditions and careful slaughter, and either provide those things for their animals or patronize farmers who do. For conscious omnivores, the problems with our disastrous system lie not in eating other animals, their status as property, nor the belief that other animals exist for us in some way, but in the *how* of animals becoming edible. Animal agriculture and agribusiness are legitimate, *so long as they are done correctly*.

In *The Ethical Meat Handbook*, butcher, homesteader, and chef Meredith Leigh articulates four principles of proper killing and eating, including "the animal enjoyed a good life ... was afforded a good death ... is butchered properly ... is cooked or preserved properly" (Leigh 2015, 1). For Joel Salatin of Polyface Farms, ethical flesh-food involves letting his chickens, cows, and pigs live in a free-range, low-tech, complementary, pastoral system (Wirzba 2007, 9). While some conscious omnivores are content to purchase items produced with these standards, others take the extra step of personally killing an animal. For this sub-movement, do-it-yourself slaughter is not only advisable; it is a crucial, if not necessary, act of individual transformation and resistance to the dominant model. Yet the idea that bleeding rightly raised animals with our own hands creates more honest flesh-food eating excludes a major consideration: the matter of power.

Core to liberationist vegan objections to industrial and humane farms alike is the use of power against vulnerable subjects who are predestined for legitimized subjugation and premeditated violence. In his critique of ethical meat, Justin Van Kleeck recalls that:

> What humans have done over thousands of years is create a situation, a system in which domesticated animals are victims by design, from birth ... humans *always* have the power, along with free reign to enact violence (of all kinds) on innocent bodies. The indelible reality of this power dynamic, which results in the killing of non-consenting individuals, also belies any notion of "ethical meat." (2017)

Animal agriculture and agribusiness, including farms with the least cruelty, depend on repressive measures. They require reproductive tyranny such as control over sexual partnerships; forced mating and insemination; extensive monitoring of fertility; and restraints against childrearing, all of which disproportionately exploits female bodies, especially mothers. Many humane farmers also employ the same techniques as industrial operations, from castration and tail docking to "grinding up male chickens at birth, using animals who have been selectively bred into shapes which cause disease, suffering and early death; forced and repeated pregnancies; separating family members for profit; and killing the animals in the exact same slaughterhouses and identically 'inhumane' conditions" (Stanescu 2013, 103).

In some humane flesh-food operations, farmers manage their soil with blood, bone, or feather meal, and fish emulsion from industrially raised animals killed in conventional slaughterhouses (Veganic Agriculture Network 2008).[2] These practices undermine the idea of happy flesh-foods.

In addition to these dissonant practices, the conceptual framework behind ethical meat is also flawed. How does one love an individual while purposely breeding her for premature annihilation? How does one appreciate the intrinsic value of another being while reducing her to consumable and sellable parts? What does it mean to honor someone while denying her desire to live? In this discourse, ideas like respect are "removed from the context of human society" and "adapted to the hierarchical human-animal relationship" (Gutjahr 2013, 382). For this reason, liberationist vegans attend to the extreme violence and suffering in animal agriculture *and* to the everyday force needed to harmonize irreconcilable interests between farmers and the farmed. Conscious omnivore storytelling minimizes the latter concern.

Tactics of evasion in conscious omnivore storytelling

Intimate, lethal violence against other animals is an act of power within a system of power. Yet in a curious twist, conscious omnivore stories make power immaterial. This erasure takes place through a variety of literary and interpretive measures that reimagine bloodshed including emphasizing affect, selective comparing and contrasting, representing victims as partners, and reframing coercion and domination.

Emphasizing affect

> I look at the blood streaming down the corrugated metal and soaking into the sawdust on the ground. *The life is in the blood.* Then I look up at Montgomery.
>
> "How are you doing, Mav?" he asks. I gri mace ...
> "It's disturbing"
> "It's supposed to be," he says. "We're not supposed to take a life and then say, Well whatever." (Mavrich 2013, 58)

On the surface, conscious omnivore storytelling is about flesh-food eaters' thoughtful, unwavering encounters with the farmed animals they consume. However, a closer read reveals these animals to be vehicles for justifying human supremacy. One tactic that hides this characteristic is preoccupation with the killer's disposition. By emphasizing affect, humane farming accounts and do-it-yourself slaughter reflections redirect questions away from the appropriateness of overpowering a vulnerable creature for profit or preference toward the appropriate feelings killers should have about such overpowering.

For example, early in *Raising Chickens for Meat*, Gwen Roland distinguishes herself as a "chicken lover" unlike those who "eat store-bought chicken" (2009, 73). Yet she does not address how she reconciles her love with choosing to buy hens who are "Genetically programmed for less than a two-month life span" or killing the hen who affectionately sat on her lap (Roland 2009, 74, 78). Similarly, in her account of beheading a humanely raised Bourbon Red turkey, journalist Ariel Kaminer admits being "Scared that I would fail—that at the moment of truth I would hesitate, and thereby hurt the bird, and scared that I would succeed and end up with blood on my hands, literally and figuratively" (2010). Yet her fears about causing harm do not prompt a change of heart, despite having a host of ethically produced, plant-based foods at her disposal.

This fixation on the killer's experience functions as a "moral catharsis" (Gutjahr 2013, 382) with the visceral response to bloodshed serving as absolution. It also arbitrarily positions conscious omnivores as exceptional killers without considering how industrial laborers feel about their actions. As a tactic of evasion, emphasizing affect also masks how conscious omnivores perpetuate the dominant human–animal binary. The result is a circular ill-logic in which killers are allowed or encouraged to kill so long as they are troubled by their killing.

Selective comparing and contrasting

> In a commercial operation, Ramona Huff says, an old cow like Frances already would have been crammed in a trailer, slaughtered, and turned into a hamburger. "That's not the deal I make with my cows." (Sinclair 2010)

Another strategy for avoiding issues of power in conscious omnivore storytelling is contrasting "compassionate" bloodshed with egregious violence. This tactic of evasion hinges also on overlooking the repressive techniques within humane operations and avoiding comparisons with more liberated spaces. For example, Huff speaks highly about letting sixteen-year-old Frances live the rest of her days, as compared to factory farms that slaughter young cows for cheap fast food. Interviewer Melissa Scott Sinclair also situates Huff between sentimental pet lovers and partakers of "shrink-wrapped steaks." The animals' "idyllic existence" and Huff's decision to spare Frances, Banana Puddin' the sow, and Pinky the "friendly as a dog" cow give the appearance that something extraordinary is taking place on the farm (Sinclair 2010). However, other elements of the story suggest otherwise.

Sinclair notes that Frances had ten calves, but does not disclose whether the cow became pregnant through natural intercourse or through a forced breeding program based on Huff's schedule and customer demands. Neither Huff nor Sinclair mention whether Frances raised all her children or if Huff stole, sold, and slaughtered them, nor do they discuss whether Frances' longevity might have as much to do with her productivity, and thus her profitability, as with Huff's "affection" (Sinclair 2010). Perhaps most glaringly, the article does not investigate Huff's rationale for refusing to kill three animals while sentencing others of their kind to die. Sinclair sidesteps this tension with the first tactic of evasion by describing Huff as "not a softie" who "does, however, get attached" (2010).

By comparing humane farming and do-it-yourself slaughter to the worst, most obvious offenders, these narratives make conscious onnivorism seem like a radical alternative to the standard system. Yet, there are other methods of agriculture and agribusiness that meet the movement's ethical concerns— *and* break with dominant ideologies about humans and other animals. Small, organic permaculture farms where animals live until they die naturally and local, organic, veganic farms that forgo using other animals are more merciful than farms predicated on killing and premature death. Selective comparing and contrasting not only masks important similarities between conventional and humane endeavors; this strategy also confuses animals living relatively unrestricted lives with animals having the freedom to exist for themselves and their communities.

Representing victims as partners

Years later, when I designed a restraint chute for holding cattle for slaughter, I was amazed that the animals would stand still and seldom resist the chute. I found that I could just ease their head and body into position by adjusting the chute … It was my job to hold the animal gently while the rabbi performed the final deed. (Grandin 1999, 161)

Huff's statement about making a deal with Frances hints at a third tactic of evasion, namely portraying animals as agreeable participants in their subjugation and demise. Temple Grandin's observation about the cows' behavior in the chute suggests submission and a willingness to die. However, what has actually transpired is that people have coaxed each cow from her home, taken her to a place she did not choose, led her into a space from which she cannot escape, and restrained her until she meets a violent end she did not anticipate. Roland's statement that her hens "were ready to lay it down by the time we picked a slaughter date" is equally deceptive (2009, 74). Not only does Roland admit that the birds were healthier than those raised in a conventional system; she clearly has total control over their fates (74).

Another method of representing victims as partners occurs when death-dealers speak *for* farmed animals. In "Would You Kill This Chicken with Your Bare Hands?," Brett Mavrich recalls his encounter with a rooster as he and other men dispatch several hens. In his words, the rooster "bebops" onto the scene, "freezes," then "saunters off as if to say, 'I'm going to pretend I didn't see that'" (2013, 58). Meanwhile, when Kaminer contacts the "next of kin" for advice on cooking her turkey, farmer Hart Perry responds, "'We heritage Bourbon reds are very tasty ... and require no seasoning since that would interfere with our inherent deliciousness'" (2010, MB1). These disturbing interpretations about violence toward farmed animals also communicate distressing messages about gendered violence more generally. Mavrich picturing the rooster as a casual passer-by during the grizzly massacre of hens and Perry bizarrely portraying himself as one of the brood while dispensing advice how to cook one of their eviscerated carcasses each makes light of the culture of brutality against female-identified persons.

What appear to be innocuous analogies between killer and cock, farmer and dead hens, reaffirm the patriarchal nature of flesh-food production, including "the end justifies the means, that the objectification of other beings is a necessary part of life, and that violence can and should be masked" (Adams 2002, 24), even when storytellers are boldfaced about their executions. Furthermore, portraits of farmed animals tacitly agreeing to die and colluding with their killers obfuscate the sinister reality that humane farms and do-it-yourself slaughter are not sites of mutual exchange. Cows, sows, and other predominantly female animals are not sacrificing themselves to nourish flesh-food eaters in a cooperative arrangement. What happens to the individuals in Grandin's chute is what happens to all animals destined for death: they are overpowered and their lives taken away, albeit under less harsh conditions. Conscious omnivorism not only supports these violations; it provides a positive framework for their continuance.

Reframing coercion and domination

> Everything at the farm, from Montgomery and me to the chicken to the land, has a Creator. And because of this, I hold no ultimate mastery over the bird I have just killed, because it wasn't mine to begin with. The hen was a gift. (Mavrich 2013, 58)

While the third tactic of evasion recreates the victim as a partner, reframing coercion and domination rehabilitates the killer by dispossessing him of his power. Even after cutting two hen's throats, a process he recounts in detail, Mavrich still concludes that he does not rule over the birds. He might have taken "its life," but the responsibility lies with a higher being (2013, 58). This perspective fits a type of Christian theology that sees creation as "an altar on which creatures are offered to each other as an expression of the Creator's self-care and provision for life" (Wirzba 2011, 112). However, it also appears in notions of a natural web of life in which all must die so all can live. The language of "gift" is instrumental here, masking the fact that some death is artificial and unnecessary. For, if animal flesh is God's or nature's offering, then slaughter is only gratitude, taking life is only receiving it, and refusing to kill and eat is sacrilege.

Another form of this fourth strategy is indiscriminately grouping the slaughter of farmed animals with other forms of killing and dying. In "What

Does It Mean to Kill Humanely?," Bryan Welch explains, "I kill some of our food animals myself, or I haul them to a slaughterhouse. I sometimes help customers kill the animals they've purchased. Once in a while I put an injured animal out of its misery" (2011, 104). Although there are social, legal, and ethical distinctions between different forms of killing when people are involved, conscious omnivore storytelling categorizes distinct types of animal death as the same. Refusing to distinguish between and make moral judgments about euthanasia or accidental killing, and slaughtering a healthy farmed infant as a matter of preference, positions the latter as inevitable. Using a passive term like "death" instead of more accurate terms like "murder" or "execution" ignores the fact that humane farming and do-it-yourself slaughter are acts of coercive systemic power.

Conscious omnivore storytelling and plantation romances

At first glance, contemporary conscious omnivore storytelling and plantation romances have little to nothing in common. One genre comprises real testimonies of killing actual farmed animals. The other consists of false accounts about slavery. However, at least two factors invite inquiry into their overlapping tactics of evasion. One consideration is the similar function of humane farming accounts and do-it-yourself reflections within the "field of social conflict" (Gutjahr 2013, 381) between liberationist and other vegans, and the role of anti-Tom novels as "a kind of literary damage control" (MacKethan 2004) for pro-slavery advocates refuting abolitionists. Each type of writing serves as defensive literature seeking to reestablish the legitimacy of systemic violence amid intensified scrutiny and charges of cruelty. They are both attempts to regain moral ground for their violent hierarchies.

Plantation romances responded to antislavery agitation with "idealistic portrayals of the master class, embellished with silent slaves, usually in the background" (MacKethan). These stories became especially prevalent after Harriet Beecher Stowe published *Uncle Tom's Cabin* (1852), selling 10,000 copies in its first week and 300,000 copies in its first year in the United States, and 1.5 million copies in one year in Great Britain (Harriet Beecher Stowe Center [a]). Stowe's two-volume narrative, which drew heavily on other abolitionist writings, dramatically advanced the antislavery cause and intensified public conversation about the institution. *Uncle Tom's Cabin* sparked numerous favorable and critical book reviews, negative newspaper campaigns and letters to editors, and inspired nationwide theater performances by people on both sides of the issue. A host of memorabilia also flooded the marketplace, including ceramics, board games, wallpapers, song sheets, and other everyday merchandise. In short order, it rapidly touched every area of society and even fanned the flames leading to Civil War (1852). In the wake of the novel's impact, slavery advocates responded with their own writing flurry. By the time union and confederate soldiers began fighting less than a decade after Stowe's publication, they had rereleased or published approximately twenty-nine anti-Tom novels disputing her work (Harriet Beecher Stowe Center [b]).

In her analysis of plantation romances and abolitionist narratives, Lucinda MacKethan describes them as "symbiotic genres" (2004). While fugitive slave narratives nevertheless reflected and sustained Southern culture, plantation romances also drew upon abolitionist themes to make their arguments. Consequently, anti-Tom novels were often "much clearer in expressing the South's anxiety about power and order than in promoting the South's confidence in its 'peculiar institution'" (2004). This dialectic also characterizes the dynamic between conscious omnivore storytelling and vegan criticism. Although humane farming accounts and do-it-yourself slaughter reflections are forms of "carnistic backlash," it is "not despite vegan advocacy, but largely because of it that such defensiveness has made its way into public discourse" (Joy 2012). These new pro-meat arguments take seriously their opponents' concerns even as their assumptions and conclusions reflect the dominant, meat-eating culture's resistance to "truly embracing a vegan ethic" (2012). This attempt to embrace liberationist ethics, while refusing their prohibition against bloodshed, is part of conscious omnivorism's internal instability.

Another reason to explore the shared ill-logics of conscious omnivore storytelling and plantation romances are the ways in which animals and slaves are defined and excluded since the introduction of race as a social construct. As Syl Ko explains, "The domain of the 'human' or 'humanity' ... is a *conceptual way to mark the province of European whiteness as the ideal way of being homo sapiens.* This means that the conceptions of 'humanity/human' and 'animality/animal' have been constructed along *racial* lines" (2017, 23, Ko's italics). Furthermore, what distinguishes the ideal from other homo sapiens is the latter's presumed closeness to the colonial, European invention of *the* animal, "with 'animals' here being a gross reduction of a vast plurality of species" (67). As the farthest away from the ideal human, animals are the most abject, and thus designated for disposal, based on their usefulness to people—and especially to *the* human. By extension, "what condemns [Black people] to our inferior status ... is not merely our racial category but *that* our racial category is marked *the most* by animality" (67). Although it is not possible in this space to fully articulate the mutually reinforcing relationship between blackness and animality, the historical and ongoing connections between these social constructs make it reasonable to cross-examine narratives that target members of these groups.

Shared tactics of evasion in plantation romances

The same tactics that appear in humane farming accounts and do-it-yourself slaughter reflections are present in at least two anti-Tom novels: Mary Henderson Eastman's *Aunt Phillis's Cabin; or Southern Life As It Is* borrowed from and responded directly to Stowe and was the best-selling book in the genre at between 20,000 and 30,000 copies (Railton n.d.[b]). Meanwhile, Caroline Lee Hentz's *The Planter's Northern Bride* took a more indirect approach, with a story of a slaveholder whose love interest with abolitionist leanings comes to appreciate plantation life. Her counter to "the dark and horrible pictures drawn of slavery and exhibited to a gazing world" (1854, iv) was republished several times in the nineteenth century.

Emphasizing affect

[Cousin Janet's] heart was full of love to all God's creatures; the servants came to her with their little ailings and grievances, and she had always a soothing remedy—some little specific for a bodily sickness, with a word of advice and kindness. (Eastman 1852, 28)

You do not speak more gently to your little sister than did [Emma Livingston] to her household slaves. I have seen her lavish the tenderest caresses on their little infants. I have seen her hang in anxious watchfulness over their sick-beds. I have seen her weep over their humble graves. (Hentz 1854, 131–132)

The tactic of emphasizing the masters' characters and their attachment to their slaves is widespread in anti-Tom novels. In *Aunt Phillis's Cabin*, the planters of Exeter are loving and devout Christians who care for their slaves like members of an extended family, and patriarch Mr. Weston is a "southern gentleman" with "a kind and charitable heart" (1852, 27). Similarly, in an aside to her primary drama, Hentz describes the Livingstons as enamored with and attentive to the needs of their slaves. Meanwhile, protagonist and slaveholder Mr. Moreland declares that "next to our own kindred, we look upon our slaves as our best friends" (1852, 24). When characters in the narratives address cruelty in the South, they identify it as an anomaly committed by people who lack good Christian morals, or who are greedy, prone to drunkenness, and mean-spirited. In this way, the good-natured dispositions of the masters are signs of slavery's integrity.

One of the foremost problems with this argument is that it is demonstrably false. Slaveholders did not "make the life of servitude … as much as possible a life of comfort and enjoyment" (Hentz 1854, 82–83) but brutally did the opposite. Yet, even if this fanciful picture of master–slave relationships had any bearing in reality, it would not erase that the institution arose from colonialism and violent removal; systematic denial of language, culture, religion, and other social organization; and extreme restrictions on individual movement and community self-determination. The master's feelings and personality traits did not alter the fact that slaves were buyable, sellable, and inheritable property and legally, socially, and politically subject to their owners' personal and economic interests. This fixation on the emotions, intentions, and manner of those who administered slavery—on the *how* of bondage—masked the multilayered consequences for those who remained vulnerable to various forms of physical, emotional, and psychological aggression. Here again, a circular ill-logic arises in which slaveholders can be slaveholders as long as they have admirable feelings about slaveholding.

Selective comparing and contrasting

Mrs. Brown's blood was up too, and she struck the poor girl in the face, and her big, hard hand was in an instant covered with blood, which spouted out from Ann's nose

"Well, Mrs. Brown, good evening," said Arthur. "I shall tell them at the South how you Northern people treat your white niggers." (Eastman 1852, 73–74)

"I've got to work for Mistress Grimby all this time … She's kept me on the go ever since the day broke, a scrubbing and scouring on all fours."

Moreland … sympathized, too, with Albert's wounded aristocracy, which had never bled so copiously before.

"My poor boy," said he, smiling at Albert's half-comic, half-rueful look, "you have not been used to such hard usage, I must acknowledge. It is well to have a taste of what the Northern bondwomen have to endure." (Hentz 1854, 90)

Anti-Tom novels also used selective comparing and contrasting with places they fashioned as more callous in order to present slavery as a suitable ethical option. They frequently compared plantation life with extreme images of deplorable labor conditions in the North, including indiscriminate violence toward Irish servants; destitute women workers; and free blacks experiencing greater hardship than they had while enslaved. Hentz also pitted a generous South against an un-Christian "native Africa" that practiced "slavery more galling," human sacrifices and cannibalism (1854, 83–84) to suggest slaves were worse off in their ancestral lands. Anti-Tom novels contrasted scenes of excessive violence and poverty with shackle-free slaves living in pleasant homes; worshipping and fellowshipping with their masters; and lazily dancing, singing, sleeping, and eating. As in conscious omnivore storytelling, these images obscured blatant and intrinsic domination between masters and slaves and foreclosed the idea of liberation elsewhere. However, this tactic did not eliminate the texts' ill-logics.

Hentz describes Mr. Moreland's devoted companion and slave Albert as "a young mulatto … handsome, golden-skinned youth" (1854, 14). In addition, Eastman's Aunt Phillis is "a tall, dignified, bright, mulatto woman" (1852, 102). Neither author clarifies whether Albert's and Phillis' racial identity resulted from consensual sexual relations—a possibility already complicated by the problem of unequal power and agency—or resulted from sexual violence. Other elements of the story also contradict the authors' assertions of peaceful coexistence. In a conversation with Mr. Weston, Aunt Phillis admits to sheltering a runaway slave in her exquisite cabin to which Mr. Weston responds with a gentle reprimand. The scene is intriguing as Eastman tries to balance assertions about Phillis' faithfulness and her master's extraordinary patience within a story of her betrayal and a slave escape. The attempted slave revolt in Hentz's story also interrupts representations of a tranquil South and inadvertently underscores turmoil within the system (MacKethan).

Representing victims as partners

My father has several times brought servants to New York, but they have never run away from him (Eastman 1852, 137).

"You know the people are all free at the North, Albert."
"Yes, master."

"And when you are there, they will very likely try to persuade you that you are free too, and tell you it is your duty to run away from me"

they couldn't come round this boy with that story; I've hearn it often enough already; I ain't afraid of anything they can say and do, to get me away from you. (Hentz 1854, 14–15)

Like conscious omnivore storytelling, plantation romances narrate the worldviews of their victims through the lens of those who wield ultimate, lethal power. In anti-Tom novels, pro-slavery advocates depicted obedient slaves, making them partners in their condition. One way of accomplishing this task was extensive dialogues in which slaves refused opportunities to become free, despite enticements by abolitionists and masters offering to let them go. Another method was to tell stories of slaves who escaped or received their freedom only to return after run-ins with unscrupulous abolitionists and malicious Northern employers. Like Grandin's chute, the victim's refusal to escape in these stories serves as evidence of approval and submission. Moreover, when authors acknowledge the victim's resistance to their condition, they claim it is rare and misguided, which reaffirms the rightness of the institution.

As with Mavrich's rooster and Perry's Bourbon Red turkey, authors of plantation romances co-opted slaves' voices in service of institutional violence and the ideologies that upheld it. Caricatures of cooperative servants detracted from the historical, legal, political, and social circumstances that made slavery possible and frustrated any real or imagined expressions of acceptance. Slaves came to the plantation through colonialism, kidnapping, and brutal voyages. They were akin to land, horses, and other property, and subject to both transfers of ownership and punishment. They endured patrols, pass systems, and laws that limited their associations and future possibilities. Plantations were not mutually beneficial communities or families, and slaves were not willing participants. Representing them as partners distorts these realities.

Reframing coercion and violence

Why, sir, do you mean to say, that the life of a slave is in the power of a master, and that he is not under the protection of our laws? (Eastman 1852, 136)

Were you to ask me if I justified the slave trade,—that traffic forced upon us, by that very British government which now taunts and upbraids us ... I would answer No! but if you mean the involuntary slavery which surrounds me and my brethren of the South, I reply, I can justify it; we had no more to do with its existence than our own. We are not responsible for it, though we are for the duties it involves, the heaviest perhaps ever imposed upon man. (Hentz 1854, 82)

Plantation romances reframed the intrinsic coercion and domination of slavery by locating power and responsibility away from masters. Mr. Moreland blames the British for creating a situation he cannot control and for which he has a "duty" to continue for mercy's sake. Similarly, when Arthur Weston discusses slavery with an abolitionist-minded colleague, he cites the South's "humane" laws as the determining factor for using violence against slaves and suggests that the few strict laws that exist resulted from "meddling, and unprincipled" abolitionists agitating slaves against their masters (Eastman 1852, 136). Arguably, the most prevalent strategy of dispossessing masters of their obvious dominating power was advocates' insistence that slavery was "authorized by God, permitted by Jesus Christ, sanctioned by the apostles, maintained by good men of all ages" (Eastman 1852, 24). That there were slaves to begin with was not the planters' doing, but rather a natural, ordained, and unalterable order. Rejecting the practice amounted to a kind of cosmic disobedience.

Here again, language plays a significant role. Just as words like "gift," "sacrifice," and "death" do not relay the forced elements of humane farms and do-it-yourself slaughter, anti-Tom references to slaves as "servants" and "dependents" reframe domination. In *The Planter's Northern Bride*, Hentz calls "the negroes of the South ... the happiest *labouring class*" (1854, vi, Hentz's italics). These and other similar terms give the impression that slavery was like other forms of employment with voluntary agreements about time and tasks. However, it was a site of daily subtle and explicit, systemic and personal subjugation, managed by the use and threat of violence and maintained by prevailing ideologies about slaves' perceived animality. While workers of the North could quit, only masters could determine when slaves were free. Even if plantation life was idyllic—and the historical record says otherwise—masters were not good stewards and slaves' lives were never fully their own.

Conclusion

As forms of defensive literature responding to threats against established and assumed systems of violence, plantation romances and conscious omnivore storytelling use a variety of tactics of evasion to paint subjugation with a benevolent, principled veneer. Noticing the overlaps in rationales for their distinct but intersecting forms of violence offers another avenue to think critically about humane farming accounts and do-it-yourself slaughter reflections, especially as they appear to offer a radically different relationship to those who have disproportionate power and uneven ethical standing. Inspecting their reasoning in light of anti-Tom novels makes conscious omnivorism's ill-logics visible. Moreover, and just as importantly, seeing how these ill-logics operate in pro-slavery and conscious omnivore arguments makes clear their danger for other overpowered groups.

Humane farming accounts and do-it-yourself slaughter reflections seem to be telling a new tale: that it is possible to use others against their own interests with compassion; to practice exploitation with love and respect; to deepen our capacity for care with experiments in unnecessary bloodshed; to honor someone's intrinsic value *and* order their lives as walking dead. However, conscious omnivorism upholds

> the mentality of domination and subjugation, of privilege and oppression ... that causes us to turn someone into something, to reduce a life to a unit of production, to erase someone's being. It is the might-makes-right mentality, which makes us feel entitled to wield complete control over the lives and deaths of those with less power—just because we can. And to feel justified in our actions, because they're only ... savages, women, animals. (Joy 2011)

The ill-logics within these stories have devastating implications—even on the most pleasant of farms—but they also have consequences beyond agricultural spaces. For this reason, those concerned with justice and liberation must recognize these tactics of evasion and confront them wherever they appear. In so doing, we can identify obstacles to appropriate care for every body, humans, and other animals alike.

Acknowledgments

Deep gratitude to Kelly Struthers Montford, Dinesh Wadiwel, and Austin Stonewall; and to Red Oak Community House in Elkhart, Indiana, and Piebird Farm Sanctuary/Vegan Farmstay in Nipissing, Ontario. Thanks also for the Race and Animals Summer Institute at Wesleyan University for creating a space to think together about some of these issues. In remembrance of Mocha and Cairo. In service to all our creaturely kin.

Notes

1　The term "vegan" is widely used to refer to a dietary choice versus its original meaning of someone who forgoes the use of nonhuman animal bodies and by-products in all areas of life. I use "liberationist vegan" to differentiate between those who adopt a plant-based diet as part of a larger commitment and set of practices toward ending individual and structural domination over other animals, and those who solely adopt a plant-based diet for ecological, health, and/or other non-animal-specific reasons. Although "ethical vegan" is the popular term to make this distinction, it is the case that eating for human health and/or the health of the planet are also forms of ethical eating. Therefore, "ethical vegan" is too vague a descriptor to distinguish it from a liberationist perspective.

2　The current food system also uses animal remains and waste in the production of vegetables, grains, and other plant-based foods. Growing interest in and development of veganic agriculture is one response to this conundrum.

References

Adams, C. J. (2002), *The Sexual Politics of Meat: A Feminist Vegetarian Critique*, Tenth Anniversary Edition, New York: Continuum.

Bouvard, V., Loomis, D., Guyton, K. Z, Grosse, Y., El Ghissassi, F., Benbrahim-Tallaa, L., Guha, N., Mattock, H., and Straif, K. (2015), "Carcinogenicity of Consumption of Red and Processed Meat," *The Lancet* 16 (16): 1599–1600.

Eastman, M. H. (1852), *Aunt Phillis's Cabin; or Southern Life as It Is*, Philadelphia, Grambo & Co. Available online: http://utc.iath.virginia.edu/proslav/eastmanhp.html (accessed September 29, 2017).

Global Data (2017), "Top Trends in Prepared Foods 2017: Exploring Trends in Meat, Fish and Seafood; Pasta, Noodles and Rice; Prepared Meals; Savory Deli Food; Soup; and Meat Substitutes," June. Available online: https://www.reportbuyer.com/product/4959853/top-trends-in-prepared-foods-2017-exploring-trends-in-meat-fish-and-seafood-pasta-noodles-and-rice-prepared-meals-savory-deli-food-soup-and-meat-substitutes.html (accessed July 29, 2017).

Grandin, T. (1999), "Slaughtering Can Be Humane," in T. L. Roleff and J. A. Hurley (eds.), *The Rights of Animals*, 159–162, San Diego: Greenhaven Press.

Gutjahr, J. (2013), "The Reintegration of Animals and Slaughter into Discourses of Meat Eating," in H. Röcklinsberg and P. Sandin (eds.), *The Ethics of Consumption*, 379–384, Wageningen: Wageningen Academic Publishers.

Harriet Beecher Stowe Center (n.d.[a]), "Her Words Changed the World." Available online: https://www.harrietbeecherstowecenter.org/harriet-beecher-stowe/her-global-impact (accessed September 23, 2017).

Harriet Beecher Stowe Center (n.d.[b]), "Stowe's Global Impact." Available online: https://www.harrietbeecherstowecenter.org/harriet-beecher-stowe/her-global-impact/(accessed September 23, 2017).

Hentz, C. L. (1854), *The Planter's Northern Bride*. Philadelphia: T. D. Peterson. Available online: http://utc.iath.virginia.edu/proslav/hentzhp.html (accessed February 4, 2018).

IPCC (2015), *Meeting Report of the Intergovernmental Panel on Climate Change Expert Meeting on Climate Change, Food, and Agriculture*, M. D. Mastrandrea, K. J. Mach, V. R. Barros, T. E. Bilir, D. J. Dokken, O. Edenhofer, C. B. Field, T. Hiraishi, S. Kadner, T. Krug, J. C. Minx, R. Pichs-Madruga, G.-K. Plattner, D. Qin, Y. Sokona, T. F. Stocker, M. Tignor (eds.), Geneva: World Meteorological Organization. Available online: https://www.ipcc.ch/pdf/supporting-material/Food-EM_MeetingReport_FINAL.pdf (accessed July 31, 2017).

Joy, M. (2011), "Carnism: Why Eating Animals Is a Social Justice Issue," *One Green Planet*, 3 November. Available online: http://www.onegreenplanet.org/lifestyle/carnism-why-eating-animals-is-a-social-justice-issue/(accessed October 11, 2017).

Joy, M. (2012), "Understanding Neocarnism: How Vegan Advocates Can Appreciate and Respond to 'Happy Meat, Locavorism', and 'Paelo Dieting'," *One Green Planet*, July 29. Available online: http://www.onegreenplanet.org/lifestyle/understanding-neocarnism/ (accessed February 19, 2018).

Kaminer, A. (2010), "The Main Course Had an Unhappy Face," *New York Times*, November 19. Available online: http://www.nytimes.com/2010/11/21/nyregion/21citycritic.html (accessed August 27, 2017).

Leigh, M. (2015), *The Ethical Meat Handbook: Complete Home Butchery, Charcuterie and Cooking for the Conscious Omnivore*. Gabriola Island: New Society Publishers.

MacKethan, L. (2004), "Plantation Romances and Slave Narratives: Symbiotic Genres," *Southern Spaces*, March 4. Available online: https://southernspaces.org/2004/plantation-romances-and-slave-narratives-symbiotic-genres (accessed October 12, 2017).

Mavrich, B. (2013), "Would You Kill This Chicken with Your Bare Hands?" *Christianity Today*, November.

Railton, S. (n.d.[a]), "Anti Uncle Tom Novels." Available online: http://utc.iath.virginia.edu/proslav/antitoms.html (accessed March 1, 2018).

Railton, S. (n.d.[b]), "Aunt Phyllis's Cabin." Available online: http://utc.iath.virginia.edu/proslav/eastmanhp.html (accessed March 1, 2018).

Roland, G. (2009), "Raising Chickens for Meat," *Mother Earth News*, June/July: 73–80.

Roth, B. (2016), "Butchering Beatitude," *The Mennonite*, September.

Sawyer, W. (2016), "Chickens, Cows and Pigs … Oh My!," *Rabobank*, August. Available online: https://research.rabobank.com/far/en/sectors/animal-protein/chicken-cows-and-pigs-oh-my.html (accessed September 20, 2017).

Sinclair, M. S. (2010), 'Facing Your Food," *Style Weekly*, December 15. Available online: https://www.styleweekly.com/richmond/facing-your-food/Content?oid=1441931 (accessed August 27, 2017).

Stanescu, V. (2013), "Why 'Loving' Animals Is Not Enough: A Response to Kathy Rudy, Locavorism, and the Marketing of 'Humane' Meat," *Journal of American Culture* 36 (2): 100–110.

Strom, S. (2017), "Americans Ate 19% Less Beef from '05 to '14, Report Says," *New York Times*, March 21. Available online: https://www.nytimes.com/2017/03/21/dining/beef-consumption-emissions.html (accessed July 29, 2017).

Van Kleeck, J. (2017), "Of Bullies and Butchers: Ethical Meat, Vegan Bullies and the Humane Myth," *Striving with Systems*, February 19. Available online: https://strivingwithsystems.com/2017/02/19/%EF%BB%BFof-bullies-and-butchers-ethical-meat-vegan-bullies-and-the-humane-myth/(accessed September 20, 2017).

Veganic Agriculture Network (2008), "Introduction to Veganics," February 25, 2008. Available online: http://www.goveganic.net/rubrique16.html (accessed September 20, 2017).

Welch, B. (2011), "What Does It Mean to Kill Humanely?" *Mother Earth News*, October/November: 103–105.

Wirzba, N. (2007), "Barnyard Dance," *Christian Century*, January 23, 2007.

Wirzba, N. (2011), *Food and Faith: A Theology of Eating*, Cambridge: Cambridge University Press.

Yu+, D. (2017), "US Dairy Milk Sales Expected to Decline until 2020, Mintel Report Shows," *Dairy Reporter*, March 16. Available online: http://www.dairyreporter.com/Markets/US-dairy-milk-sales-expected-to-decline-until-2020-report-shows (accessed July 29, 2017).

Chapter 4

ERUPT THE SILENCE

Hayley Singer

Here is a woman. Her husband describes her as timid, of middling height, plain as a black shoe. She has, he thinks, a "passive personality" (Kang 2007, 3). She exhibits neither "freshness nor charm, or anything especially refined" and is "completely unremarkable in everyway" (2007). One morning, this woman stands in the dim light of her fridge, bare foot and silent as a ghost. And then she empties it. Gone is the beef for shabu-shabu. Gone is the belly of pork and the sides of black beef shin. Gone are the squid, the sliced eel, the frozen dumplings, the eggs, and the milk. Gone, all gone, into a black rubbish bag. Her husband is distressed: is she crazy? All night she has had dreams, painfully vivid visions, "A long bamboo stick strung with great blood-red gashes of meat, blood still dripping down" (12). Strange, horrible, uncanny nightmares that chart the ugly legacies of carnist violence. Soon, she won't be able to sleep for fear of more dreams. "Dreams of murder. Murderer or murdered ... hazy distinctions, boundaries wearing thing" (28). Her husband thinks she is self-centered: "So you're saying that from now on, there'll be no meat in this house?" (13). He is yet to abuse her. Her family is worried. Vegetarianism is unnatural. Her father is ashamed. He is yet to abuse her.

The woman, the vegetarian, is not given space to tell her version of the story. Like so many female figures in literary history (think Bertha Rochester, that iconic mad woman in the attic; think Elizabeth Costello, whose transgressions of the generic conventions of the academic humanities leave her words to slide into a kind of obscurity), her story is taken from her. It is denied, censored, refused, put into quarantine. As novelist and essayist Porochista Khakpour writes, it is only through "something approaching a post-language state" (2016, 2) that is used to recount Yeong-hye's dreams can a reader come to know the horror that haunts this woman, the protagonist of Han Kang's novel *The Vegetarian*. But, her silence is not inactive. It is not docile. It shifts cultural dynamics; it creates gender, species, and genre trouble.

She is meat. She. And so speaking meat she lives the
silence. Pitch-black. Silence.

Literatures that attend to the intersections of speciesism, racism, classism, sexism, able-bodism, and ageism listen for, and document, silences: their causes and their effects. They do so in order to address exclusion and inclusion, power and powerlessness, voice and voicelessness: who can speak, who is listened to, who listens, who is authorized to speak, and who takes for themselves the authority to speak?

In her essay "A Short History of Silence" (2017), Rebecca Solnit writes that silence, which is a tool of oppression and therefore different from quietude, still needs to be interrogated and annihilated. There is a world of difference, she writes, between quiet and silence. Quiet is something that might be sought after, while silence may be imposed: "Quiet is to noise as silence is to communication" (18). Solnit charts the ways silence precedes, supports, and results from violence. Her essay moves through a nuanced landscape in which different codes and kinds of silence—those forged through shame, politeness, fear, indifference, the legal status of powerlessness, ridicule, ostracism, and trauma—concrete into cultural habitats in which speaking up and calling attention to silence and violence are acts of empathy. She traces the lineage from second- to third-wave feminist literatures, and literary analysis, to understand the way women have turned to silence as (and turned silence into) a serious topic of concern. From Rachel Carson's *Silent Spring* (1962) to Deborah Levy's *Diary of a Steak* (1997) and Alexis Wright's *The Swan Book* (2013), intersectional literatures have shown that liberation from silence is a storytelling process and a fundamental part of articulating the knots of violence in which all earth-bound critters are living and dying. Importantly, these texts also recognize that silence is a politically potent form of poetic expression.

Cartographies of silence have also been traced and tracked in critical ecological, feminist-animal, and vegan-feminist scholarship, including Carol J. Adams' germinal works *The Sexual Politics of Meat* (1990) and *The Pornography of Meat* (2003), Val Plumwood's deeply influential study *Feminism and the Mastery of Nature* (1993), Lori Gruen's *Entangled Empathy: An Alternative Ethic for Our Relationships with Animals* (2015), and studies like Susan Fraiman's 'Pussy Panic versus Liking Animals: Tracking Gender in Animal Studies' (2012). There are so many kinds of silence to contemplate in these texts: the silence of the "absent referent" (Adams 1990, 5–6); silences produced by Western philosophy's discourse of rationalism (Plumwood 1993); silences created by philosophical abstraction, individualism, and impartiality (Gruen 2015); and the bibliographic silences produced by overlooking the rich bodies of work written by women within the fields of critical ecological and animal studies (Fraiman).

In carnist systems of knowledge there is a strong desire for anti-carnist silence. Silence remains the persistent (enforced) condition of most animals in contemporary culture. As Elizabeth Costello (the protagonist of J. M. Coetzee's

(2004) highly influential novel of the same name) notes, "Animals have only silence left with which to confront us" (70). Silence, in this context, might be understood as critiquing, condemning, bearing witness to, and acting as testimony to the experience of erasure that colonizes the lives and deaths of so many animals.[1]

> *The thing that hurts is my. All those remnants. I don't know why. Sticks in the chest. That lump.*

How to work with silence as a form of poetic expression? Can silence be used as a storytelling technique? As a writer, I wonder how I can conspire with silence to read it as a disruptive narrative tactic. Can it offer new forms of imagining, resisting, revolting, living, dying, and telling stories of the myriad lives and deaths that make up this damaged planet? What if silence is not docile but citational—pointing to something that is purposefully kept off the cultural scene? Can silence, as a citational practice, act as a form of resistance—one that teaches something about the characteristics of the stories given cultural primacy? Can silence help stories unfurl the ongoing destruction of carnism and its interconnected systems of violence?

For Rebecca Solnit, having a voice and being heard is part of living as "a human being with full membership" (24). But, what if a person does not want to be folded into the field of humanity? What if, as Carol J. Adams wonders, one is invested in exploring what it means for humans to recognize themselves more fully as "animal beings" (2006, 120)? And what if a person has more vegetal desires—desires to phytomorphize? What kinds of stories will they need to tell?

"A free person tells her own story," Rebecca Solnit writes (2017, 19). But I want to think about those who are not freed by this act. Those people who are rendered socially unintelligible, inaudible, dysfunctional, untrustworthy, and even indigestible because of the stories they need to tell. That is the story of Yeong-hye.

Originally published in South Korea as three novellas—*The Vegetarian, The Mongolian Mark,* and *Flaming Trees—The Vegetarian* burrows into the social, cultural, and political fibers of carnist violence. As the novel opens, the reader is confronted with Yeong-hye as a young suburban woman who has had a sudden experience of disgust in relation to meat. She has had dreams, particularly visceral dreams of slit flesh, pools of blood, butchered bodies, screams, breathlessness, and darkness. These dreams have formed a council of images that haunt her; they are the specters of all the animals she has eaten in her life.

In her attempts to step away from carnist frameworks of thought and action, Yeong-hye renounces human and nonhuman animal characteristics. She feels her body transform into a vegetal being. She responds to light and shadow with phytomorphic desires, she feels roots grow from her hands and flowers emerge from her vulva. Coming to exist within root networks that link veins to vines, birthmarks to blooms, and blood to chlorophyll, Yeong-hye shrugs "off flesh like a snake shedding its skin" (170). Within the novel this botanical turn poses significant challenges to carnist and humanist identities. It allows Han Kang to complicate ideas of embodiment, communication, kinship, nourishment, life, death, and time.

> *Four juddering legs. The rope. The blood. Its dead eyes.*
> *I remember. I did scoop up a mouthful.*

In an interview published in the online journal *Lit Hub* (Patrick 2016), Han Kang says that *The Vegetarian* questions the relentlessness of human violence and the (im)possibilities of innocence from this condition of violence. The drama of *The Vegetarian* emerges as the novel traces the "undoing" of Yeong-hye, a woman who cannot be anything but the antithesis of the self that is demanded by carnism. At a family lunch, Yeong-hye refuses the family law that demands she eat meat.

> My mother-in-law brought in dishes of stir-fried beef, sweet and sour pork, steamed chicken, and octopus noodles arranging them on the table in front of my wife.
>   "This whole vegetarian business stops right now," she said. "This one, and this, and this—hurry up and eat them. How could you have got into this wretched state when there's not a thing in the world you can't eat?" (Kang 2007, 36)

In this passage, and those that follow below, Yeong-hye lands the force of a disobedient vegan silence into a world where what is audible/intelligible is authored in the image of carnivorous humanity. "The children were staring wide-eyed at my wife. She turned her blank gaze on her family, as if she couldn't fathom the reason for all the sudden fuss" (Kang 36).

Yeong-hye's gaze is not so blank, it acts as a spectator of the spectral, she holds her eyes firmly on what, in the contemporary carnist moment, is occluded and occulted: the ghosts of animals turned into meat. Those ghosts are gathered up into Yeong-hye's silent stare and she brings them with her to the family table. Bringing occulted others to the table, being *cum panis*[2] with the ghosts of industrial slaughter, threatens the ordinary carnist subject with "invasion" or "contamination" by these dead others. Her silence becomes an expression, and an enactment, of trouble. The living infects the dead. Yeong-hye feels herself to be infected with the dead. She is full-up with death and can eat no more. As Yeong-hye's husband narrates, "This time, my mother-in-law picked up some sweet and sour pork with her chopsticks and thrust it right up in front of my wife's mouth, saying, 'Here. Come on, hurry up and eat.' Mouth closed, my wife stared at her mother as though entirely ignorant of the rules of etiquette" (37).

For Solnit making new stories means breaking silences, but as Han Kang shows in *The Vegetarian*, silence can be a powerful tool in breaking old stories. French psychoanalyst and literary theorist Luce Irigaray (1980) once wrote, "If we keep on speaking the same language together, we're going to reproduce the same history. Begin the same old stories all over again. Don't you think so?" (205). Yes, of course. To say what remains unsaid, to chart what is politically and culturally different, is to wander into the dark spots of one's era.

In his essay "What Is Contemporary" (2009), Italian philosopher Giorgio Agamben asks what it means to look into the dark spots of one's era, to perceive the "special darkness" of one's time, to catch sight of the obscure things (the things kept off the cultural scene) and never cease to engage them (44). Such a task is, he writes, an imaginative and a physical act: one's eyes must be "struck by the beam of darkness that comes from his own time" (45). Agamben's thinking holds important threads for my own contemplation of silence in *The Vegetarian*, of silence as an act of vegan disobedience not a lapse into madness. Agamben recognizes that those who let their eyes latch onto the darkness of their times will, simultaneously, be taken up by that darkness: submerged, subsumed, engulfed, overwhelmed, inundated, swamped. While Agamben's thinking is focused on the practice of certain writers, his argument holds true for the work of Han Kang and her character, Yeong-hye. Yeong-hye is held by the nightmares that she holds in her mind. She touches the ghosts that reach out to touch her. This reciprocity leaves deep traces in her identity. They trace the outline of a story she will no longer uphold.

> "Open your mouth right now. You don't like it? Well, try this instead then." She tried the same thing with stir-fried beef, and when my wife kept her mouth shut just as before, set the beef down and picked up some dressed oysters. "Haven't you liked these since you were little? You used to want to eat them all the time." (Kang, 37)

Yeong-hye's silence breaks with the familiarity of her family's carnist narratives. In so many ways, breaking with carnist narratives makes space for breaking bread with other "companion species of terra" (Haraway 2016, 55). Breaking, failing, refusing, ignoring, wilfully forgetting to follow pathways of thought, speech, and action: these are some of the ways to unmake carnist narratives. I take these as modes of thinking, knowing, and writing that might allow for an understanding, and further development, of anti-carnist ways of being in the world. These might be ways to stray from the bounds of carnist knowledge, while keeping in touch with the ways animals live, how they die, and why they haunt.

Wilfully forgetting carnist rules of etiquette becomes a form of embodied resistance.[3] Silent resistance turns Yeong-hye's body into outlaw, a pirate. This kind of resistance is interruptive of "normal" modes of functioning. It is also citational as it points to what is kept off the cultural scene. This kind of embodied resistance is a practice of forgetting. Yeong-hye forgets the customs that uphold carnist manners and patterns of thought, patterns that snuff out potentials for new knowledge formations. Her resistance pokes holes in carnism's toxic discourses, predictable and instituted stories—stories that function in carceral ways because they attempt to shut up the wild possibilities for ways of living and dying that may be offered by anti-carnist imaginings.

> My father-in-law took up a pair of chopsticks. He used them to pick up a piece of sweet and sour pork and stood tall in front of my wife, who turned away.
>
> ...
>
> "Eat it! Listen to what your father's telling you and eat. Everything I say is for your own good. So why act like this if it makes you ill?" (Kang, 38)

To forget is to break with lineage. To break with lineage is to shift the dynamics of times-yet-to-come by refusing to enact what is expected to come next. The carnist mentality that presides over *The Vegetarian* situates "the vegan" (Yong-hye) as an incomprehensible, irrational, and even an impossible identity because she does not live out carnist expectations.

> With one hand my wife pushed away his chopsticks, which were shaking silently in empty space.
>
> "Father, I don't eat meat."
>
> In an instant, his flat palm cleaved the empty space. My wife cupped her cheek in her hand. ... I approached my wife hesitantly. He'd hit her so hard that the blood showed through the skin of her cheek. Her breathing was ragged, and it seemed that her composure had finally been shattered. (Kang, 38–39)

In the moments that follow, Yeong-hye grabs a little fruit knife and cuts open one of her wrists. Why are some traditions more important than some lives?

In these quoted scenes, carnist patriarchy and carnist nationalism patronize: they argue, they bully, they hit. Carnism (that nasty little tyrant) has powerful tantrums.

Throughout *The Vegetarian*, Yeong-hye's silence "outs" two of carnism's most significant pathological symptoms: (1) the utter conviction in the right and "rationality" of using all animals and ecologies for one's own gains, desires, and luxuries; and (2) the hysterical focus on meat as a singularly exceptional source of sustenance and identity. By doing this, *The Vegetarian* uses silence as a narrative strategy to lay bare (and amplify) the mechanisms by which carnism relentlessly attempts to fix women, vegans, animals, and plants in the place of absolute Others, by projecting onto them an incomprehensibility and unintelligibility constituted as the refuse of human, masculine, and carnist "rationality."

The first red is blood. The dream is flesh. Bury the dream.
But first. Erupt the silence.

How to distinguish between different kinds of silence: remembrance, stealth, protest, (self) censorship, guilt, obedience, complicity, support, or the slack-jawed void of ignorance? Is Yeong-hey's silence a kind of revolutionary (to turn, to roll back) silence in that it unravels the way carnist discourses twist the realities of mass slaughter into narratives of health care?

> "You waste away after months without meat, it seems," says Yeong-hye's mother to Yeong-hye's husband, "so ... eat this together, the two of you. It's black goat. ... Try feeding it to Yeong-hye, just tell her it's herbal medicine. I put a load of medicinal stuff in it to mask the smell." (Kang, 45)

In Tillie Olsen's essay "Silences in Literature," published in *Silences* (1978), Olsen writes that the history of literature is riddled with natural and unnatural silences. Unnatural silences include deletions, omissions, abandonments, and refutation of subject matter. Unnatural silences are present everywhere in carnist systems of representation and reality. As Adams has written in her deeply influential book *The Sexual Politics of Meat* (1990), there is the silence of the "absent-referent"—those animals who have been "cut off" from their own bodies through slaughter (21). Then there are those smothered bodies, smothered lives, smothered deaths, and smothered stories that James Stanescu narrates in his article, "Species Trouble: Judith Butler, Mourning and the Precarious Lives of Animals" (2012). Try mourning those "'nameless and faceless deaths' that come to constitute our world" (579) and you will find yourself neck deep in a political practice that opens you up to the possibility of losing your social intelligibility, credibility, and legibility. Lose intelligibility, Stanescu writes, and you become spectacle, spectral, and a spectee. These three conditions converge within Yeong-hye's body as she stops eating meat, then stops eating altogether. As her body begins to eat itself her anti-carnist narrative is twisted into a discourse of madness: schizophrenia, catatonia, anorexia. Thinking with these familiar narratives the medical staff caring for Yeong-hye subject her to the kind of treatment in which the forceful preservation of a certain kind of life takes precedence over asking the question (and then listening for an answer!): What do anti-carnists want? Yeong-hye's doctor explains to her sister: "But we're still not sure why exactly it is that Kim Yeong-hye is refusing to eat, and none of the medicines we've given her seem to have had any effect" (141).

Schizophrenia, catatonia, anorexia: these three conditions used by family and doctors to colonize Yeong-hye's silent story, which continues (nonetheless) to gather human-animal-plant bodies up into the drama of living and dying together. In doing this, Yeong-hye's silence breathes with (conspires) gender, genre, and species trouble. Neither her body, nor her silence will go docilely into quarantine, nor will they be readjusted to fit a carnist perspective of life.

Her silence is not a narrative wasteland/dead zone/vacuum but a position of poise—a narrative position that remains alert to the possibility that there are myriad ways to live outside the single-minded, and ultimately deadly, narrative of carnism.

> *One time. I want to shout. Just one time. To pitch.*
> *To yell. To howl. Yes, perhaps that could work.*

The Vegetarian explores enmeshments of the more-than-human with the inhuman, human-as-humus, human-as-animal, human-as-meat, human-as-vegetal, human-as-spectee, and human-as-fiction. When, in the latter parts of the novel, Yeong-hye feels herself transforming into a plant, it comes as her final (and most extreme) attempt to renounce all kinds of carnist violence. As she mentally metamorphoses, Yeong-hye's desires to live a plant-based and plant-bound life are interpreted as signifiers of delusion.

This novel can be read as a tale for what Donna Haraway calls the Chthulucene. Like Haraway's cyborg from "A Cyborg Manifesto" (1991), the Chthulucene is a figure for thinking with, a way to frame narratives of life and death. In her recent book *Staying with the Trouble: Making Kin in the Chthulucene*, Haraway describes the Chthulucene as a figure of reality and of imagining, of science and of fiction. The concept of the Chthulucene is taken from, but not rooted in, H.P. Lovecraft's "misogynist racial nightmare monster Cthulhu" (101). Rather, the Chthulucene describes an era that stands within and alongside (while challenging) the Anthropocene, Capitalocene, and Plantationocene. It is an era that acknowledges the "myriad intra-active entities-in-assemblages" (1991) that constitute life on earth. The Chthulucene is an age that thrives on stories, on speculation, speculative feminisms, and speculative fabulation.

Haraway's figure of the Chthulucene functions in a practical way for writers and other kinds of thinkers and activists, because it invites us to reframe the stories we tell so that we might learn the art of staying with complex (and often troubling) questions of multispecies justice in intersectional ways.

To tell stories in and of the Chthulucene is to tell stories rooted in the soil, in the earth's humus. It is to tell stories of earthly presence and entanglement. Yeong-hye's silence contains, and even speaks, the precarious, unfinished, and deadly conditions in which animals caught in the systems of industrial animal agriculture are forced to live. Yeong-hye's body is at stake to sheep and lambs and cows and chickens and fish. Her silence speaks the madness of not taking the lives and deaths of animals born into industrial agriculture seriously. Haunted by meat, she turns to cellulose in search of a new narrative frame, another mode of being. The roots that Yeong-hye imagines to be growing out of her hands act as feelers and, as Haraway points out, the Latin word for "feeler" is *tentaculum*, from which the word *tentacle* emerges (31).

Using imaginative finger-roots to feel her way beyond carnism, Yeong-hye threads herself to a network of stories in which women and animals witness oppression, alienation, and disfiguration with poetic and politically potent forms of silence.

Kang's complex, poetic use of silence leads me to read Yeong-hye's character as one rooted in the historic figure of the "snaky Medusa," a figure Haraway takes as one of her many "chthonic ones" (71). Chthonic ones embody the Chthulucene as their stories, and their concerns are rooted in the earth. They are critters with many legs, critters who spin webs and stories. They are those like "Naga, Gaia, Tangaroa, Terra, Haniyasu-hime, Spider Woman, Pachamama, Oya, Gorgo, Raven, A'akuluujjusi, and many more" (Haraway 1991). They are those who "romp in multicritter humus but have no truck with sky-gazing Homo" (2). They are immunized to human exceptionalism. They act as speakers for the dead and because of this they are available for multispecies storytelling.

Just like Yeong-hye, chthonic ones take risks in how they live, what they imagine, whose stories they hear, and whose points of view they hold alongside their own. It is my reading that Yeong-hye's dreams of meat and murder, as well as her phytomorphic desires, make her a chthonic one. In this way, Yeong-hye can be read as a literary affiliate of the Medusa whose stare petrifies because it conveys all the horrors of disfigurement.

The Italian philosopher and feminist thinker Adriana Cavarero unpicks the etymology of horror to discover its links with repugnance and the physiological state of paralysis. In her book *Horrorism: Naming Contemporary Violence* (2009), Cavarero ties the power of the mythical Medusa's silent stare to this etymological nexus. For Caverero, Medusa's stare petrifies because it speaks of the kinds of violence not merely content to kill but driven to "destroy the uniqueness of the body, tearing at its constitutive vulnerability" (8). Yeong-hye's silent stare speaks of the destruction of uniqueness that is just one of the dictates organizing the logic of the abattoir. So it is that I take Medusa and Yeong-hye as literary oddkin. They are, to use a phrase from queer-trans-feminist-literary scholar Jack Halberstam, "fugitive knowers" (2011, 8)—ones who unleash forms of knowing, valuing, and mourning that are predominantly ignored, lost, or forgotten. Yeong-hye's silent stare looks directly at connections that are feared. And she catches herself in the silent stare of the animals who haunt her. Yeong-hye is horrified, but she cannot look away. Rather, she turns from the narratives, identities, and actions that produce, legitimate, normalize, and maintain such horrors: carnism. Yeong-hye no longer wants to know the world through the narrative of carnism. For her, carnism has become utterly unthinkable.

In her introduction to a collection of essays *Mourning Animals* (2016), Margo DeMello writes that while today there are "countless options for dealing with the death of a companion animal" (xvii) there are fewer ways to publicly acknowledge and mourn the vast numbers of animals killed in processes of industrial meat, dairy and egg production, scientific testing, or as a result of the unpredictable and "slow violence" (Nixon 2013, 2) produced by environmental catastrophes of climate change, toxic drifts, oil spills, deforestation, and the environmental aftermaths of war. Tracing the "animal fragments" in Butler's work, Stanescu says that to ignore the deaths of those millions of "nameless and faceless" animals, whose exploitation is woven into the fabric of everyday life, is to ignore the death of part of ourselves (569). Precisely, we lose that part of ourselves that is made up of the myriad connections we have with those others. "To disavow the death of another achieves two kinds of isolation," he writes: "it cedes others into social unintelligibility as well as making one unintelligible to themselves" (569). For Stanescu, the practice of creating "species trouble" draws on and remains intimate with gender trouble. In *Precarious Life: The Powers of Mourning and Violence* (2004), Butler writes, "I propose to consider a dimension of political life that has to do with our exposure to violence and our complicity in it, with our vulnerability to loss and the task of mourning that follows, and with finding a basis for community in these conditions" (19). Stanescu applies this idea to the practice of publicly mourning animals. Continuing to show connections between gender and species trouble, Stanescu describes the way rituals of public grief can take deaths that are "invisibilized" and amplify them in order to allow them the weight, reality, and voice they deserve (578–579). The first step in this practice of mourning is to risk coherency. To make gender trouble, to make species trouble, is to make genre trouble. Yeong-hye does handstands in the muddle of all this trouble.

Yeong-hye's silence never loses touch with questions of ethics, justice, or response-ability. It opens up, and speaks of, the gaps between how things are and how they might be. Silence in *The Vegetarian* calls me to uncomfortable sensibilities because it is, at once, a tool of oppression *and* a mode of poetic expression. It is a force of annihilation just as it can be used to critique oppression and act as a practice of liberation. Silence can erupt from indifference *and* from care shocked into stupefaction. Silence offers space to hear murmurings, muted cries, and mutterings. Silence provides space for ghosts to talk and be heard. Across *The Vegetarian* Yeong-hye's silence turns haptic, it unravels into/across/ beneath carnist structures and identities. Attempting to discipline Yeong-hye's silence is the drama that consumes the narrative. Her silence has voice; it is a kind of rhetoric. It erupts, disrupts, speaks up, and reveals. In doing so, it makes space for a new narrative world.

Notes

1   As Rachel Carson has shown in no uncertain terms, silence is just one of the many effects that result from the introduction of poisons (used as herbicides and pesticides) into ecosystems, which result in the unintended killing of numerous insects, birds, fish, cows, sheep, and even some people.

2 I take this term from Donna J. Haraway who uses it to describe companion species meeting, being, and eating at table together. See chapter 2 "Tentacular Thinking: Anthropocene, Capitalocene, Chthulucene" in *Staying with the Trouble: Making Kin in the Chthulucene*, 55.

3 In the essay "The Everyday Resistance of Vegetarianism," which comes from the edited collection *Embodied Resistance* (2011), Samantha Kwan and Louise Marie Roth theorize the way acting out certain bodily practices, and abstaining from "normalizing" bodily practices, can form modes of resistance to and interruption of institutional power. Here, I draw on their ideas of "counter-hegemonic embodiment" and use it to describe the bodily resistance that comes from acting out veganism.

References

Adams, C. J. (1990), *The Sexual Politics of Meat: A Feminist-Vegetarian Critical Theory*, New York: Continuum.

Adams, C. J. (2006), "An Animal Manifesto: Gender, Identity, and Vegan-Feminism in the Twenty-First Century," *Parallax* 38: 120–128.

Agamben, G. (2009), "What Is Contemporary?" in D. Kishik and S. Pedatella (trans.), *What Is an Apparatus? And Other Essays*, 39–54, Stanford, CA: Stanford University Press.

Butler, J. (2004), *Precarious Life: The Powers of Mourning and Violence*, New York: Verso.

Carson, R. (1962), *Silent Spring*, Boston: Houghton Mifflin.

Cavarero, A. (2009), *Horrorism: Naming Contemporary Violence*, New York: Columbia University Press.

Coetzee, J. M. (2004), *Elizabeth Costello*, London: Vintage Books.

DeMello, M. (2016), "'Who Goes to the Rainbow Bridge?' Conceptions of the Afterlife for Non-Human Animals," in M. DeMello (ed.), *Mourning Animals: Rituals and Practices Surrounding Animal Death*, xvii–xxvi, East Lansing: Michigan State University Press.

Gruen, L. (2015), *Entangled Empathy: An Alternative Ethic for Our Relationships with Animals*, Brooklyn: Lantern Books.

Halberstam, J. (2011), "Shadow Feminisms: Queer Negativity and Radical Passivity," in *The Queer Art of Failure*, 123–146, Durham: Duke University Press.

Haraway, D. J. (2016), *Staying with the Trouble: Making Kin in the Chthulucene*, Durham: Duke University Press.

Irigaray, L. (1980), "When Our Lips Speak Together," in C. Burke (trans.), *This Sex Which Is Not One*, 205–218. New York: Cornell University.

Kang, H. (2007), *The Vegetarian*, London: Portobello books.

Khakpour, P. (2016), "*The Vegetarian* by Han Kang," *The New York Times*, February 2. Available online: https://www.nytimes.com/2016/02/07/books/review/the-vegetarian-by-han-kang.html (accessed August 5, 2016).

Kwan, S. and Roth, L. M. (2011), "The Everyday Resistance of Vegetarianism," in C. Bobel and S. Kwan (eds.), *Embodied Resistance: Challenging the Norms, Breaking the Rules*, 186–196, Nashville: Vanderbilt University Press.

Levy, D. (1997), *Diary of a Steak*, London: Book Works.

Nixon, R. (2013), *Slow Violence and the Environmentalism of the Poor*, Cambridge: Harvard University Press.

Olsen, T. (1978), "Silences in Literature," in S. Fisher Finkin (ed.), *Silences*, New York: The Feminist Press at CUNY.

Patrick, B. (2016), "Han Kang on Violence, Beauty, and the (Im)possibility of Innocence," February 12. Available online: http://lithub.com/han-kang-on-violence-beauty-and-the-impossibility-of-innocence/ (accessed August 5, 2016).

Plumwood, V. (1993), *Feminism and the Mastery of Nature*, London: Routledge.

Stanescu, J. (2012), "Species Trouble: Judith Butler, Mourning, and the Precarious Lives of Animals," *Hypatia: A Journal of Feminist Philosophy* 27 (3): 567–582.

Wright, A. (2013), *The Swan Book*, Artarmon: Giramondo.

Chapter 5

THE LONELINESS AND MADNESS OF WITNESSING: REFLECTIONS FROM A VEGAN FEMINIST KILLJOY

Kathryn Gillespie

"How do you cope?" One of the students asked, intently—searchingly.

I froze. It was a reasonable question. Especially so, considering I had been invited into this space—a student-led undergraduate course on self-care and trauma in social justice-oriented work—to talk about my own experiences related to the theme for the week: *What to do when people don't care about what you care about.* My mind raced: *They're asking for strategies. What strategies for coping or self-care do I have? What can I offer?* And then my mind went blank, panicking. I felt a quiver move through my body. I felt my chin begin to tremble. *How do I cope? Is that what I'm doing? Coping?* I felt my face start to flush.

How do I cope? How do I cope? Everyone in the room was looking at me, waiting. I wondered what coping looks like and if that's what it looked like I was doing. I wondered if looking like I was coping was why I had been asked to come in to talk to the class. *I'm not coping,* I thought. *I haven't been coping for a long time.* I felt my vision go blurry as my eyes filled with tears. *Oh no,* I thought frantically. *Don't cry. Not here. Not now. Get it together, Gillespie!* I looked up and they were still waiting.

Tears catching in my throat, "I'm not coping," I said. "I haven't been coping for a long time." And then I broke down, mortified, but without any control over the full-bodied, undignified sobs that cut through the silent room.

"Oh god! Oh my god!" one of the students blurted out, too loudly for the space. "Are you OK? Are you OK? Oh my god!" and then, to her classmates: "What do we do? What should we do? Oh god!"

Haley Bosco Doyle, who was facilitating the class and who had invited me to attend, turned to the other student and said, "It's OK. Let her be." Then Haley turned to me and put her hand on my shoulder gently and waited. They all waited while I covered my face with my hands and sobbed.

When I had finally gathered myself together, we talked about how out-of-place tears can feel in an academic classroom. How classrooms, and academic spaces more generally, are framed as spaces of intellectual engagement and not emotional outpourings. I talked to them about my mortification at crying in front of them, and about how it might be useful to try to reframe how I was thinking about those tears. What if they weren't something to be ashamed of; what if, instead, we could think about crying—*a teacher crying*—as a subversive act: a rejection of masculinist, rationalist notions of the academy as an unemotional space, or academics as unemotional beings. I left the classroom that day, emotionally undone.

I'd like to say that this undoing in front of a roomful of students was a breaking point or a turning point of some kind, but it wasn't. It is a memorable moment of rupture, a moment where the damage done by my experiences witnessing violence bubbled up to the surface to remind me that I was not OK. That in witnessing, studying, teaching, and writing about violence against animals, I will probably, in some ways, never be OK again.

This was early 2016. Four years earlier, in 2012, I had completed the fieldwork for my dissertation on the lives of cows in the Pacific Northwestern United States dairy industry. My intention was to document as best I could the embodied experiences of animals—cows, bulls, calves—raised for dairy production. This work involved entering spaces where the visceral experience of suffering and, often, dying animals was made mundane and routine through the abstracting effects of the commodification process (Gillespie 2016, Gillespie 2018). Farmed animal auction yards were one of my primary research sites. In these spaces, I witnessed countless instances of animal suffering and violence not typically viewed as violence: cows collapsing and dying in the rear holding pens and in the auction rings; animals being jabbed with electric prods, leaping forward to avoid the shocks; cows and calves sold apart from one another as they bellowed for each other from across the auction yard; cows crammed into pens so tightly they couldn't turn around; day-old calves with their umbilical cords still attached selling for as little as $15, or lying crumpled and weak in the holding pens. There were cows in all stages of physical health: some were robust and still viable as commodity producers; many others were emaciated and barely able to walk; many limped severely, their hides covered with open wounds and udders red with infection. Intentionally witnessing the trauma animals experience in farming is, for the witness, a psychically overwhelming onslaught of visceral suffering—an exposure to secondary trauma.

At the same time, the speed, efficiency, and everydayness with which animals are sold through the auction ring has the effect of obscuring or making mundane the violence experienced by each singular cow passing through the ring. To repeatedly reject this abstracting effect and focus in on each body, each life, and each face requires you to become the witness, the killjoy. Witnessing the suffering of animals in spaces of routine animal use is an act of embodying and inhabiting what Sara Ahmed has identified as *the feminist killjoy*. Ahmed writes, "When we make violence manifest, a violence that is reproduced by not being made manifest, we will be assigned as killjoys. It is because of what she reveals that a killjoy becomes a killjoy in the first place" (Ahmed 2017, 256). The killjoy is a witness. The killjoy bears witness to acts of violence and injustice and documents them as such—in memory, in speech, in writing, in the acts of recollection that mark and politicize this violence.

I have written elsewhere about my feelings of grief and mourning in my role witnessing the suffering experienced by cows in the auction yard (Gillespie 2016). And these experiences and moments of seeing this violence *as violence* were, indeed, overwhelmed by lasting grief and sadness, but they were also marked by profound loneliness and by feelings of madness. To intentionally focus on each individual suffering being—to remind myself again and again that *this is the cost of commodifying life*—made me feel like witnessing this violence for another second would push me "over the edge" into a kind of abyss, a place of being unhinged, inconsolable, and permanently changed.

And yet, I found myself moving forward. I didn't witness this violence for just another second, but for too many more seconds to count—too many animal lives to count or remember or grieve. And in order to witness, I found a way to contain these feelings of madness—not eliminate them, but contain them, and push them down, in order to survive. Ahmed tells us that "[the feminist killjoy] comes to exist as a figure, a way of containing damage, because she speaks about damage." I watched as these animals passed by and I contained these feelings of grief and feelings of being driven to the edge of sanity. I felt anger and disbelief and shame and desperation and I contained those, too. I moved on to the next field site. I wrote my fieldnotes. I finished my fieldwork. I taught courses on violence against animals. I prepared to write my dissertation. I defended the dissertation and finished the Ph.D. I rewrote the research in book form, all the while working hard to contain the damage, because it *was* damage—layers of damage: the internalization of the damage humans cause to the animals they farm and the damage to the witness of this damage as a result of witnessing.

Witnessing can easily drive you to madness or exacerbate mental health issues already there. It can carve out new psychological and emotional wounds. And it can open old ones, worrying at the tender scars until suddenly what you thought was healed, or at least contained, is open again, raw, festering, and exposed to the world. For me, the act of witnessing has done both. Witnessing animal suffering, for me, has exacerbated these familiar affective specters of madness, deepening them and fleshing out their contours. But they have also cut new wounds, laying bare new realities of violence, suffering, and apathy.

There are layers of damage in witnessing. There is the act of witnessing that, in spaces of animal agriculture like the auction yard, left me with overwhelming feelings of despair, rage, shame, and grief. There's something about the scale of animal agriculture—the 50 billion land animals killed each year for food— that is incomprehensible. You try to comprehend it, to contain it—you witness one animal suffering and dying before your eyes and you know that that one is billions—but the scale of it is uncontainable. You can't grieve for 50 billion animals at once and contain that grief. You try. You fail. Your mind works hard to abstract from the animals before you—to blur your vision with tears, to set a ringing in your ears to obscure the sounds of the suffering—to protect itself and to protect your heart. You don't see or remember the next few animals passing before you. You feel shame that you can't focus on and remember each one. You try harder. You find yourself abstracting again. You exhaust yourself with the effort.

And then you notice that no one around you is having this experience: they are laughing, bidding, snacking, chatting with friends and family. This pushes these feelings of madness further. You return to your life—changed, undone—to find that most people don't know about, care about, think about the suffering of animals at all. And this is another layering of the madness of witnessing. You've just witnessed this violence and you stand, outside of (and always within) spaces of violence against animals, always a witness, a killjoy, a reminder that this violence is happening all the time. You find yourself alone. Ahmed writes:

> A killjoy might first recognize herself in that feeling of loneliness: of being cut off from others, from how they assemble around happiness. She knows, because she has been there: to be unseated by the tables of happiness can be to find yourself in that shadowy place, to find yourself alone, on your own ... The costs of killing joy are high; this figure herself is a cost (not to agree with someone as killing the joy of something). (Ahmed 2017, 259)

I am lucky to have a close handful of family, friends, and colleagues who understand or who have been willing to try to understand violence against nonhuman animals *as violence*. Some of these people already cared about violence against animals and others have come to care about it because it is something I care about. Within this circle, I am also lucky to have a partner who is enormously supportive and has engaged in this journey of learning about violence against animals with me, and a close friend and colleague who has accompanied me into some of these spaces and engaged in endless conversations about this work (see Lopez and Gillespie 2016).

Still, most people don't understand nor do they bother to try to understand what animals experience—indeed, many people take offense at the thought that farmed animals' lives and experiences matter in any consequential way. Or, others think the matter of caring about animals is cute or laughable or merely a manifestation of privilege. Working on and caring about violence against animals frequently renders one socially unintelligible. Social unintelligibility, as James Stanescu writes, is "a failure of recognition by others, a failure to code as reality what you know reality to be. It is an erasure of existence, an erasure of sense, and an erasure of relations. To have your grief for one you care about rendered unintelligible does not invite simple ridicule; it invites melancholia and madness" (Stanescu 2012, 579). And so, sometimes it is easier—less painful—to try to cope, or to pretend to cope. At least in trying to cope, it is only the pain of what you have witnessed that you must carry. To expose yourself as not coping, to try to talk about the pain of what you have witnessed with people who don't see it as intelligible adds a layer of isolation and a feeling of going mad that only works to amplify the damage.

But there's also a profound loneliness to looking like you're coping when you're not. Thinking back to the theme of that day's class—*what to do when people don't care about what you care about*—there's a profound loneliness to caring about things that other people don't care about. There's a profound loneliness when people don't care about what you care about. Sometimes the impacts of people not caring about what I care about are so much that there have been times I've wished desperately that I could just not be the killjoy doing the work of killing joy for even a few moments or a day.

These moments of disavowal have occurred in my personal life in many different contexts and they regularly happen in my academic life. For instance, I have had many of these moments while on the academic job market, which for the killjoy can be its own kind of mental health nightmare. For any academic, the experience of going on the job market is fraught, stressful, and sometimes traumatizing. Ann Cvetokovich describes the experience:

> Huge uncertainties about the future—Is my intellectual work any good? Will I get a job? Where will I be living? Is this really what I want to do?— coalesce around an endless array of tiny tasks and decisions about everything from which font to use on a CV to how to describe one's dissertation, which represents years of work, in a single paragraph that will capture the attention of an unknown reader. (Cvetkovich 2012, 30–31)

This is compounded by the stress of waiting to hear (often waiting for a rejection that hiring committees might never even bother to send), the stress of interviewing and campus visits (if you're among the very few who land one), the unsettling reality of uprooting your life and moving to a new place, making new friends and colleagues, and so on. Any academic who's been on the job market in recent years will have their own set of stories from their own personal hell of trying to get an academic job.

All of this is magnified and transfigured for the killjoy. For the killjoy whose witnessing work is also about killing joy, the job market is a field of emotional and psychological political landmines. On one side, you're tasked with selling your research over and over again in neatly curated descriptions, all the while pretending that this research hasn't driven you to madness. You try to write about it in applications or talk about it in interviews in ways that don't betray that you have been completely undone by it, that just thinking about it makes you want to close yourself into a dark closet and cry.

On the other, you have to package your research just right—so that it is not too political, not too obviously killing joy, to get in the door. And this process is made even worse if your work is about killing joy in relation to people's everyday practices of consumption (like uncovering the violence of dairy production). When confronted with my first campus visit for a feminist geography job, I was advised to frame my work through its explicit contributions to feminist theory; I developed a job talk that highlighted these contributions. Still, my dissertation was about violence against cows in routine practices of agricultural production; there was no getting around or concealing this fact, and this violence was framed as a distinctly feminist issue. As everyone settled into their seats and I began my talk, I watched with a sinking feeling as a platter of cheese and crackers was passed through the audience. *Oh no.* I thought. *I'm fucked.* As the roomful of academics (many of whom were self-described feminists) nibbled on their cheese, I detailed the gendered violence experienced by cows raised for that cheese. Ahmed writes:

> Sometimes, when [the killjoy] appears on the horizon of our consciousness, it can be a moment of despair. You don't always want her to appear even when you see yourself in her appearance. You might say to her: not here, not now … You might think, you might feel: I can't afford to be her right now. You might think, you might feel: she would cost me too much right now. (Ahmed 2017, 171–172)

Why did it have to be cheese? I thought. *I just want a job. Can't I just not be the killjoy for once?* The committee didn't bother to notify me of my rejection for that job.

Colleagues regularly remind me that rejection on the job market is not personal, but when your work is about witnessing and making legible routinely obscured forms of violence, it is nearly impossible to not take it personally insofar as I regularly feel that I failed the animals whose lives and deaths populate my research. It is not that these animals would care whether or not I had gotten a job; rather, rejection represents a feeling that I didn't do enough to communicate the urgency and importance of attending to violence against them. And this is part of being the killjoy more generally—the way killing joy, witnessing, making political becomes so thoroughly entangled with the killjoy's own life that sometimes I think, at my very core, I have been reconfigured and defined by these forms of violence and others' responses to them. Internalizing this violence and its social unintelligibility marks the body and mind in profound ways and I am still working to understand how this occurs.

The first time I read Johanna Hedva's "Sick Woman Theory," I felt a part of the loneliness I was feeling shatter: she was speaking about the damage done by encountering, being exposed to, being haunted by normalized forms of violence. She wasn't talking about violence against animals, but this didn't matter; her words validated and authenticated the psychic and embodied effects of being the killjoy, the misfit, the outcast, trying to love and care and live in this world. She writes: "the body and mind are sensitive and reactive to regimes of oppression—particularly our current regime of neoliberal, white-supremacist, imperial-capitalist, cis-hetero-patriarchy. It is that all of our bodies and minds carry the historical trauma of this, that it is *the world itself* that is making and keeping us sick" (Hedva 2016).

Because of the privileged space I inhabit in the world (white, physically able-bodied, cis-gender, born in the United States, human), I am in many ways insulated from feeling the full effects of many direct forms of structural violence. And yet, if we take Hedva's words seriously—that "it is that all of our bodies and minds carry the historical trauma" of our current and historically embedded regimes, then the damage caused by intentional (and sometimes unintentional) witnessing of violence (against animals and others) refracts the echoes of these traumas and violences and animates them anew.

Whether we identify as the killjoy, the witness, or not, the impacts of the violent regimes we live in take their toll—and this occurs radically differentially and unequally depending on levels of vulnerability and precarity. But becoming the killjoy is also another kind of taking-on of damage and committing to disrupt, communicate, and (hopefully) transform, through the act of taking on that damage and refusing to obscure it.

Reading Hedva's work alongside Ahmed's exploration of containment and damage in the killjoy's reality, I understand us all to be containers of damage. "To be a container of damage is to be a damaged container; a leaky container. The feminist killjoy is a leaky container. She is right there; there she is, all teary, what a mess" (Ahmed 2017, 171). If *"the world itself"* (Hedva 2016) causes us all damage, then we all become containers of damage and damaged containers, absorbing the damage caused by the current and historical violent regimes we inhabit. To be considered *normal, healthy,* and *happy*—and to make others feel comfortable and happy—it is expected that we be good at making it look like the damage isn't there. Feeling blue or under the weather are acceptable temporary states of being in response to certain stimuli (a sad or tragic event, a virulent cold virus), but to remain in these states of revealing the damage (the clinically depressed, the chronically ill, the neuro-atypical, the perpetual witness) is to become Ahmed's leaky container; it is to become the killjoy (*"there she is, all teary, what a mess"*); it is to become the figure that ruptures the fiction that everyone (no, *any*one) can be healthy or happy when the world is so thoroughly imbricated with these regimes of oppression that take shape and form in our bodies and minds.

It is in this realization that I am learning, in an unexpected way, to celebrate the killjoy in all her teary, leaky, witnessing mess and to find gratitude in the damage she contains and expels—I've come to find a strange sense of warmth and familiarity in the feelings of madness that are called forth and energized by her presence. Because I see these experiences of witnessing violence against animals as a memorable point of rupture—a place of springing a leak in the container—I sometimes wonder what it would be like to go back to before I chose to expose myself (through witnessing, writing, and teaching) to the ubiquitous world of violence against animals. Becoming the killjoy is painful. It adds an additional psychic burden to the deeply buried wounds that may already be there. In these flickers of thinking about the past, I feel nostalgic for a time when these wounds were not so numerous and when those that were already there were more deeply buried. I feel nostalgic for a time when I was not so actively and relentlessly the leaky container of the killjoy. But this seductive nostalgia is hard to sustain because I wouldn't choose to not have witnessed these things and I wouldn't choose to not be changed by them.

More than that, though, even if I wanted to engage in this undoing of damage, I couldn't. I was already a container of damage, only, perhaps, I was a little less leaky. The surfacing and layering on of the damage of witnessing has shaped and continues to shape who I am in my body and mind and heart. Eli Clare, writing on the politics of cure and pushing back against ableist notions of normative forms of embodiment and neurotypicality, asks: "Can any of us move our bodies back in time, undo the lessons learned, the knowledge gained, the scars acquired? The desire for restoration, the return to a bodily past—whether shaped by actual history, imagination, or the vice grip of *normal* and *natural*—is complex" (Clare 2014, 211). In much of his work, Clare refers to the body-mind as a rejection of the mind–body dualism and as an acknowledgment that who we are is a complex and inextricably entangled fusion of embodiment, thought, and feeling. Moving our body-minds back in the temporal arc of our lives is an impossibility. We are shaped so thoroughly and "brilliantly imperfectly" (Clare 2017) by how we arrive in the world, by how our experiences reorient and transform and undo and remake our body-minds. Who would we be without this madness, this feeling, these politicized manifestations of witnessing and care?

Acknowledgments

Many thanks to Lori Gruen and Fiona Probyn-Rapsey for their encouragement and editorial oversight and curating such an interesting and important project as *Animaladies*. Special thanks also to Haley Bosco Doyle for creating such a warm, thoughtful, and inclusive space for so many activists and academics to think through the emotional and psychic tolls of doing this work. To Eric Haberman, my ongoing gratitude for all of the support, love, and understanding given so freely and generously. Finally, the writing of this chapter was made possible by the wonderful time and space provided by the Wesleyan Animal Studies Postdoctoral Fellowship.

References

Ahmed, S. (2017), *Living a Feminist Life*, Durham, NC: Duke University Press.
Clare, E. (2014), "Meditations on Natural Worlds, Disabled Bodies, and a Politics of Cure," in S. Iovino and S. Oppermann (eds.), *Material Ecocriticism*, 204–218, Bloomington: Indiana University Press.

Clare, E. (2017), *Brilliant Imperfection: Grappling with Cure*, Durham, NC: Duke University Press.

Cvetkovich, A. (2012), *Depression: A Public Feeling*, Durham, NC: Duke University Press.

Gillespie, K. (2016), "Witnessing Animal Others: Bearing Witness, Grief, and the Political Function of Emotion," *Hypatia* 31 (3): 572–588.

Gillespie, K. (2018), *The Cow with Ear Tag #1389*, Chicago: University of Chicago Press.

Hedva, J. (2016), "Sick Woman Theory," *Mask Magazine*, January. Available online: http://maskmagazine.com/not-again/struggle/sick-woman-theory (accessed May 2, 2017).

Lopez, P. J. and Gillespie, K. (2016), "A Love Story: For 'Buddy System' Research in the Academy," *Gender, Place, and Culture* 23 (12): 1689–1700.

Stanescu, J. (2012), "Species Trouble: Judith Butler, Mourning, and the Precarious Lives of Animals," *Hypatia* 27 (3): 567–582.

Part II

DISABILITY

Chapter 6

ABLEISM, SPECIESISM, ANIMALS, AND AUTISM: THE DEVALUATION OF INTERSPECIES FRIENDSHIPS

Hannah Monroe

Closeness with nonhuman animals is an often discussed aspect of autistic people's experiences.[1] It is present in autobiographies by autistic people and has been written about in research into autistic people's experiences (Solomon 2015).[2] The prevalence of this topic is especially significant given that autistic people are often thought of as not being able to relate to others; but their closeness with animals disrupts this view (Bergenmar et al. 2015; Solomon 2015). Preconceptions and stereotypes surrounding gender, autism, and empathy further complicate understanding the experiences of autistic women in particular. I am connected to this research and the autistic community as both a scholar and an activist for neurodiversity, as part of a movement that questions the construction of autism as pathology within medicalizing discourse. For my master's thesis I interviewed four autistic young adults about their experiences of being autistic in discursive contexts, and I will draw on my research for a part of this chapter.

In this chapter I explore ideas of interspecies closeness between autistic people and nonhuman animals. I will first give an overview of connected oppressions between nonhuman animals and autistic people, including the wider community of disabled people. Then I will discuss notions of empathy as they are connected to gender and autism, and the impact these ideas may have on autistic women. Next, I will explore the ways connections with nonhuman animals are experienced. It is important to avoid generalizing aspects of people's experiences, so I want to address the level to which stereotypes about autistic people complicate the experience of connection to animals, as well as nature, in the lives of many autistic people. Not all autistic people experience these connections to animals and nature. Nonhuman animals and autistic people can be allies for each other, and these connections can be a site of solidarity in working against speciesist and ableist oppression. Relatedly, I will discuss how feeling closer to nonhuman animals than others of one's own species is pathologized as a symptom of autism and show how this distorts an important connection and the solidarity found in it. Lastly, through a discussion of Temple

Grandin's approach, I will explore how this feeling of closeness can sometimes be harmful for nonhuman animals.

Connected oppressions

Disability scholars critique the overly rationalist way in which animals are discussed in animal rights discourse, noting that these kinds of arguments are rooted in ableist and speciesist thinking and ultimately do not help animals.[3] These traditional animal rights arguments privilege "rationality" as a basis for beings deserving rights, seemingly leaving out nonhuman animals and people who think and behave in non-normative, non-neurotypical, and non-conforming ways. As Taylor asks, "isn't it ableist to devalue animals because of what abilities they do or do not have?" (2017, 57). Furthermore, this argument upholds dualist binaries of rational and irrational that perpetuate ableist and speciesist oppression.

Taylor suggests that speciesism is supported and constructed by ableism, which justifies viewing abilities associated with neurotypical humans as more valuable than those associated with nonhuman animals (58–59). Gruen also problematizes the argument for caring about and respecting animals because they "share many of the qualities that we admire in ourselves and to which we attach moral significance" (2015, 17). This argument, perhaps unwittingly, promotes anthropocentrism, ignoring the existence of important traits outside of "typical" human experience (24–25).

Salomon argues that privileging the ability to reason is connected to the oppression of autistic people, because neurotypical ways of thinking are held as a standard for others to be measured against. He calls the privileging of these ways of thinking "neurotypicalism" and notes that "this privileging of vermal reasoning over other forms of reasoning not only invalidates and makes suspect autist insights, but neurotypicalism also invalidates and makes suspect animal intelligence" (2010, 48–49). Vermal reasoning is "reasoning that relies heavily on the brain's vermis" (48). He categorizes nonhuman animals as also non-neurotypical (50). Because autistic people and nonhuman animals both may be nonconforming to neurotypical ways of thinking, privileging vermal reasoning is speciesist and ableist.

Taylor asserts that the reasons nonhuman animals are regarded as so different from humans, and thus denied personhood, are deeply rooted in ableism. Nonhuman animals are often regarded as different for not aligning with neurotypical norms of intelligence. Thus the denial of personhood to nonhuman animals is connected to ableism. She asserts "ableism allows us to view human abilities as unquestionably superior to animal abilities" (2017, 58–59). Similarly, Bergenmar et al. argue that the concept "human" is constructed in such a way that we have to meet certain criteria to reach it (2015, 203). They argue that those regarded as not fitting these criteria are animalized or medicalized, and they seek to explore these criteria for "humanness" in regard to norms of emotion and social interaction (204). They draw on Davidson and Smith's work on the connections with nonhuman animals described by autistic people's autobiographies, and focus on how these autobiographies address the constructed "qualifications for obtaining 'humanness'" (207–208). By writing about interspecies experiences, and talking about ways they actually feel like nonhuman animals, autistic authors "question the concept of being 'human'" and in the process disrupt binary dualisms of human and nonhuman animals (208, 210). For example, one author cited by Bergenmar et al., Johansson, describes viewing nonhuman animals, the natural world, and humans all as equal (211). Another author, Prince-Hughes, wrote that she "grew to understand and identify with the gorillas, the way other human people did *not* identify with them" (209). In these authors' feelings of connection to nonhuman animals and the natural world, they disrupt normative constructions of personhood.

One of the central ideas in disability theory is that our mental and physical abilities are not what make us valuable and deserving of dignity and respect (Taylor 2017, 57). Disability activism urges us to see different ways of being in the world as valuable and positive, and Taylor and Salomon argue that this view is essential for animal advocacy as well (Taylor 2017, 60; Salomon 2010, 47–49). The devaluation of autistic people's ways of living and thinking is an example of a prejudice that needs to be overcome. Instead of determining value based on how close abilities are to those of able-bodied and neurotypical humans, we should value each other's ways of being in the world as valuable in and of themselves. Taylor asserts, "disability activists do not argue that disabled individuals are valuable *despite* our disabilities; rather, value lies in the very variation of embodiment, cognition, and experience that disability encompasses" (60). We need diversity of expressions and experiences because, as we are interdependent, we all contribute to helping each other in different ways. Furthermore, constructed norms represent few people's actual experiences. Embracing and valuing different ways of being can create a more equitable society for all.

Empathy, gender, and autism

The clinical view of autism describes autistic people as lacking the ability to empathize, not being able to see how someone else might be feeling and respond to that state appropriately (Davidson and Smith 2012, 262; Bergenmar et al. 2015, 204–205). Davidson and Smith draw from evidence of autistic people's close bonds with nonhuman animals and nature to question whether this is really because autistic people lack empathy or whether they just have difficulty interacting socially with other people (2012, 262). In autobiographies and narratives, autistic people express that they experience "difficulties in communicating or understanding emotions" (267). They have trouble communicating emotionally with other people, but this does not affect their empathy. In fact, autistic people describe experiencing emotions to an extreme intensity, called "hyper-emotionality," which is one reason why connecting with other people can be so difficult (Davidson and Smith, 268; Bergenmar et al., 206). Experiencing emotions from connecting with other people can be overwhelming, while those experiences in nature or with nonhuman animals are less intense (Davidson and Smith, 268).

Assumptions about empathy and autism are rooted in normative ideas about gender. Solomon describes Baron-Cohen's claim that autistic people have mental traits more stereotypically associated with "males." Solomon challenges this view, writing "such a dichotomously gendered view, however, may obscure the experiential complexity of affective experiences in ASD" (2015, 339). Solomon notes that autistic boys have been found to be social and empathetic with nonhuman animals as well. She also describes how this stereotypical view of autism is changing, with recognition of the diversity in autistic people's experiences and abilities taking its place (339).

Baron-Cohen also claims that women are more empathetic than men, but Jordynn Jack points out how the evidence for this is based on studies that were biased due to problems in their methodology, where researchers did not account for participants' knowledge of the study: when women participated in studies that they knew were about empathy, they would often demonstrate greater empathy than men, but this has to do with gender norms and women feeling pressure to be more empathetic. Studies that avoided this problem did not find gender differences in empathic responses (2014, 136). It is significant that women do feel such pressure to be empathetic, especially as autistic people are stereotyped as having less empathy. This places autistic women between two contrasting stereotypes. They may experience social pressures to be more empathetic in accordance with gender norms while also being assumed to have less empathy because they are labeled as autistic.

These converging expectations and stereotypes demonstrate the need for an awareness of diversity and intersecting identities, such as race, gender, and sexuality, in clinical ideas around autism. The assumption that autistic people experience less empathy than neurotypical people, and particularly its reliance on gendered stereotypes, is demonstrative of the widely held belief that autism is more common among men than women. Gendered stereotypes about autism render autistic women and their experiences invisible. Feeling that one's identity is invisible in both cultural and clinical views of autism surely affects autistic women's sense of selves. These gendered and ableist stereotypes about autistic women and empathy are disrupted by the close connections some autistic people have with nonhuman animals.

Emotional connections

Close connections with nonhuman animals, as well as nature, play a significant role in autobiographies of autistic authors and are often described as having the emotional depth and closeness of human friendships. Such connections disrupt the idea that autistic people do not have these emotional experiences (Davidson and Smith, 260). Solomon also explains that as autism "has been conceptualized as a disorder of affect, it casts doubt upon the very ability of people with ASD to experience attachment, care, and the intersubjective understandings of others" (327). But the level of connection in autistic people's interactions with nonhuman animals challenges this idea about autism (327). It becomes clear that this idea about autistic people's social and communicative abilities is more complicated than is predominantly thought. One child in Solomon's research, for example, Kid, is described as not getting along well with other children, but as loving nonhuman animals. Solomon explores the way Kid shows care and empathy for animals, disrupting the assumption that autistic people do not experience emotional connections (333).

Davidson and Smith also bring up the idea of "a person who claims to *feel more like* an animal, or even a place, than they feel like other people" (266–277, 271). Bergenmar et al. describe the narrative of a Swedish autistic person, Brändemo, who says he is like his pet rabbit because they share traits associated with his autism. He and the rabbit share similar sensory and social experiences (209). It is particularly interesting that he expresses this connection in terms of feeling that he is like the rabbit, rather than as a sense of merely feeling understood. This contrasts other accounts of connections with animals written by autistic people, in which they emphasize feeling better understood by animals in comparison to other humans and the relative ease of reading and understanding animals' emotions (Davidson and Smith, 270, 272).

In my research, I asked four participants who they felt understood by, and if this included nonhuman animals.[4] Participants did not describe feeling understood by animals, but some instead talked about feeling like or relating to animals. Perhaps feeling like a nonhuman animal is another kind of emotional connection that is helpful for autistic people and nonhuman animals in a similar way. For example, Mikkel says, "I don't know if they understand me, but I feel like very related to cats. I feel I understand cats better … People, they say they're mysterious animals. And then don't really get them. But I just see them as kind of independent animals. And I kind of like that."[5] He does not feel like cats understand him, but seems to see parts of his own identity in cats and feels like he understands them while other people do not. This connection to cats could still be affirming for him, even if he does not feel understood by them.

I found others expressed relating to nonhuman animals, but not feeling understood or liked by them, while some did not feel this kind of connection but still expressed a fondness for animals and had pets. For example, Grace feels understood by animals because they communicate in ways she can understand. However, she does not feel that any animal she has interacted with has really liked her. This is another distinction, where being able to communicate is not the same as being liked. Being able to communicate with and understand animals in some ways is not the same as having a close emotional bond. This is interesting in that it provides a different narrative, demonstrating diversity in autistic people's experiences. Now I will explore the ways these connections, for the people who do experience them, can be a source of alliance and solidarity.

Possibilities for alliance and solidarity

There are concrete ways that autistic people and nonhuman animals can be allies for each other and be in solidarity. Taylor argues that acknowledging interdependence is critical to disability and animal justice. However, for disabled people and nonhuman animals, being cared for has often been paradoxically linked with oppression. For disabled people the experience of being cared for often means having less independence and agency, and for nonhuman animals it is often connected to their violent exploitation (205–206). In our culture, being dependent is also constructed as negative, with disabled people and domesticated nonhuman animals being viewed as burdensome.

Disrupting this idea, Taylor argues that all of us are dependent on each other in many ways (209–210). Everyone in our society relies on others in all aspects of our lives, such as people who build and maintain infrastructure in our neighborhoods, schools, and cities. We rely on medical professionals to help take care of our health. As we go about our day, we are reliant on people who grow food for us as well as those who serve us in restaurants and cafes. Interdependence is deeply embedded in culture and society. Referencing the feminist ethic of care and Gruen's work on empathy, Taylor argues interdependence should require that we care for each other in non-oppressive ways, and that would include listening to those we care about to understand how to care for them and what they want. She says we need to "start listening to what those who need care are communicating about their own lives, feelings, and the care they are receiving" (218). She calls for a need to recognize the ways we are interdependent with nonhuman animals, including how we can help and benefit each other in ways that do not undermine agency and rights (218). An example of this would be an autistic person gaining validation and emotional support from a nonhuman animal they live with and caring for their companion animal in return. Taylor shares Gruen's account of entangled empathy that stresses the need to be aware of the ways we are both the same and different. Though we are interconnected and interdependent, we also have our own selves. Gruen argues that "value dualisms" in which we see one as superior to another are harmful, but this does not mean we have to disregard differences between ourselves and others. Importantly, Gruen explains that for marginalized people having a "self-identity" is something they have had to claim for themselves in the face of oppression, and thus can be empowering (Gruen in Taylor 2017, 62). This is crucial in reference to autism and identity.

Davidson and Smith describe how connections with nonhuman animals and nature can offer "respite from the disruptive, intrusive, and communicatively overburdened social world" (261). In a social world outside of human interaction, these individuals find connections that are more accommodating of their communication styles (269). For example, in Brändemo's experience as well as that of Prince-Hughes', it is easier to be social with nonhuman animals as their ways of living and communicating are more accommodating of autistic differences (Bergenmar et al., 209). Brändemo and Prince-Hughes describe actually learning and speaking the language of the animals they felt connected to, gorillas for Prince-Hughes and rabbits for Brändemo. In this way, they could communicate with nonhuman animals even though it bothered humans when they made their communicative noises (213).

The nonhuman animals provide a kind of protection and relief from problems that autistic people face in human society, and this "protected time" is often among nature and nonhuman animals. Autistic people do not avoid connection in this safe space, but rather want to connect with different species and the natural world (Davidson and Smith, 269). Davidson and Smith express the sense of refuge autistic people can find in nature and among nonhuman animals, describing that this provides a space for "ASD authors to simply be and feel themselves, comfortably and even pleasurably, however divergent from the norm their feelings are taken to be by clinicians" (269). This really describes how liberatory these connections can be. One of the authors Bergenmar et al. write about, Johansson, says it is easier to relate to nonhuman animals and the natural world because they do not expect anything of her, whereas humans do and often what they expect is harmful for her (211). Solomon suggests a reason the autistic children in her research find it easier to be social with nonhuman animals could be that they have more agency in these interactions, as human contexts for autistic children are very controlled (338). In addition, "the animal gaze can be felt to affirm rather than weaken a sense of self and self-worth" and provide "gentle recognition and (almost unconditional) acceptance" (Davidson and Smith, 274). This description of the interactions between nonhuman animals and autistic people reveals how these connections help with navigating an ableist society. Autistic people's identities and ways of being are oppressed to the point of being detrimental to self-worth. For this reason especially, finding other beings who will be accepting and affirming is critically important.

Expanding this discussion to disabled people's experiences more broadly, I would like to mention Taylor's story about her service dog, Bailey, who is also disabled both physically and mentally, having experienced trauma prior to being adopted (220, 223). Her description of how Bailey helps her is significant, especially in connection to descriptions of the ways nonhuman animals help autistic people in social ways. She describes how as a physically disabled person, being in social spaces is difficult for her because of the ableist ways people interact with her. Taylor writes, "One of the most powerful services animals can provide is a certain kind of social ease, mediating between their human companions and an ableist world" (221). In this way Bailey has been an ally for her in coping with ableist oppression. As Bailey became physically disabled, Taylor recognized that he is her service animal, especially in social ways, and simultaneously she is his service human, helping accommodate his physical and mental disabilities. In this way, she shows us an example of liberating interdependence and interspecies solidarity. While these connections are clearly affirming and empowering for many people, they are also frequently pathologized and devalued through a combination of speciesism and ableism.

Pathologizing of interspecies connections

The devaluation of connections autistic people experience with nonhuman animals shows how they are seen as less socially significant than others of our own species. I believe this is an example of speciesism and objectification, as nonhuman animals are not recognized as social beings in community with us. Within the dominant discourse, animals are constituted as objects and therefore friendships with them are only seen as a substitution, and not as real or meaningful. The objectification of nonhuman animals and the construction of them as inferior to humans is closely connected to the ableist dismissal of the social connections some autistic people may have with nonhuman animals.

One can see from autistic people's narratives about communicating and emotionally connecting with nonhuman animals that these experiences are significant for them. Bergenmar et al. assert that even though these narratives disrupt the clinical view that autistic people cannot experience emotional connection with others, these experiences are often overlooked because they exist outside of neurotypical norms of communication and empathy (214). In this way the meaningful social bonds autistic people form with nonhuman animals are made invisible and subsequently devalued.

Salomon describes how "neurotypical society is suspicious of and threatened by the special relationship autists have with nonhuman animals" and argues this demonstrates the speciesism within our culture (63). Davidson and Smith describe narrative experiences of feeling very close to nonhuman animals, in this case cats, and how the individuals wanted to be friends with cats, but were discouraged by their parents who wanted them to socialize with other children instead (270). This effort of parents to push children toward human friendships rather than interspecies ones demonstrates the pathologizing of these friendships as a symptom of autism and developmentally "abnormal." This is not how parents want their children to behave and socialize.

It seems that in a culture that so devalues and oppresses nonhuman animals, caring about them and feeling close to them is seen as abnormal. The construction of caring about animals as mental illness was more direct in the late 1800s when doctors labeled animal rights activists with "zoophilpsychosis," a constructed mental illness that supposedly made people care too much about animals. Women were diagnosed with this more often (Davidson and Smith, 151). This is addressed by Fraser and Taylor in this volume. The construction of a mental illness to pathologize caring about animals is interesting to think about from a post-structural disability studies and neurodiversity perspective, as this highlights social justice issues rooted in diagnostic labels and who receives them.

While autistic people's close connections with nonhuman animals are often pathologized, these connections are at other times seen as a way for autistic people to become more social with other humans. Davidson and Smith assert that social understanding and feeling comfortable around nonhuman animals can help autistic people develop the same feelings toward other humans (275). Similarly, Solomon asserts these connections with nonhuman animals are important to autistic people's being "social and inter-subjective" (325). She compares two situations that autistic children are involved in, one child is in a psychological interview and the other is participating in an activity with a therapy animal as well as interacting with other animals (325). Referring to autistic and neurotypical children, she writes, "in families with pets in the United States, parents overwhelmingly agree that the animal is beneficial to the children's development" (327). Nonhuman animals are viewed as helping children develop social skills, such as having empathy, which they can then bring to their relationships with other people (327).

Describing the empathetic interactions one child, Kid, has with nonhuman animals, Solomon asserts "sociality may be less a quality of the individual and more a capacity realized through certain kinds of social interactions that are unremarkable and ordinary only within certain interactional substrates, and entirely invisible and unrealizable in others" (336). As mentioned earlier, Solomon demonstrates the way the sociality and intersubjectivity of the children in her research are overlooked, particularly because their connections to nonhuman animals are ignored. She also describes nonhuman animals helping these children develop the "capacity" for this (Jack 2014). However, Solomon does not imply as strongly as Davidson and Smith that this capacity could and should transfer over to interactions and sociality with other humans, as she focuses on the importance of recognizing how autistic children may experience sociality differently around nonhuman animals (229).

I would question what this argument implies about how we view nonhuman animals. It seems that animals may be thought of as a means to an end in attempts to encourage autistic people to be social in ways they may not want to be. Even though these authors describe connections to nonhuman animals that are not pathologized outright, they are still viewed as specifically related to the individuals' autism, and as being positive in their usefulness for potentially changing autistic people's behavior. Wolfe calls this "seeing the non-human animal as merely a prop or tool for allowing the disabled to be mainstreamed into liberal society and its values" (2008, 122). Instead Wolfe suggests a model for supporting and helping each other similar to Taylor's previously mentioned views about interspecies solidarity and interdependence. Aside from the ways autistic people's understanding and connections with nonhuman animals are pathologized, the cultural assumptions about autistic people's ability to do this can also be harmful by perpetuating speciesist oppression.

For example, in discussions of autistic people's feelings of closeness with nonhuman animals, Temple Grandin's narratives are frequently mentioned.[6] As I have talked about how the connections autistic people may have with nonhuman animals can be places for solidarity and alliance, it is important to address Grandin's harmful empathy, given that her form of connection leads to harm and death. She talks about her ability to understand nonhuman animals, even saying "that she feels more *like* these animals—not simply that she likes or feels more *for* them—than people" (Grandin in Davidson and Smith, 271). Grandin's description of how she experiences empathy connects to Gruen's discussion of entangled empathy and critique of Grandin. Gruen describes entangled empathy as being able to understand how another being feels in a situation and being inspired to come to their aid. Her description of empathy and how it leads people to ethical actions questions whether approaches to empathy such as Grandin's should truly be considered empathetic (2018). Davidson and Smith suggest that it may actually be because of her statements and discussions of empathy with nonhuman animals in her work that Grandin is seen as an "expert" in designing slaughterhouses (273). Lion points out that as Grandin describes having a connection to nonhuman animals to the extent that she can communicate with them, she has become viewed as a spokesperson for them. This constructs the death of animals in the slaughterhouses she designs as positive for them, because she is seen as having a unique knowledge of their perspectives and wants (2016). Davidson and Smith rightly state "that this might seem a strange mode of employment for someone claiming to empathetically understand how cattle 'feel'" (273). I find it especially problematic that Grandin uses what could be a site of alliance with nonhuman animals as a way to prove her ability to design ways to kill them.

In referencing how nonhuman animals and disabled people can be allies as discussed by Taylor, it is important here to reiterate that this means a reciprocal relationship where nonhuman animals and disabled people can help each other. In Grandin's case, animals help her feel understood, but her work contributes to their violent oppression. Welfarist animal activists would argue she is making their deaths more humane, but as Lion asserts, by promoting a narrative of killing animals in a supposedly empathetic and understanding way, she is only supporting the idea of humane meat, which helps the industry far more than the animals being exploited.

Through discussing notions of closeness between autistic people and nonhuman animals, in this chapter, I have attempted to highlight solidarity while also acknowledging potential issues of ableism, speciesism, and gendered stereotypes. It is important to acknowledge the complexity of this topic while looking at the potential for alliance between autistic people and nonhuman animals.

Notes

1   I use the term "autistic people," which is called identity-first language, and is preferred by most autistic people. This term affirms that autism is part of a person's identity instead of an illness.
2   For autobiographies, see Bergenmar et al. (2015); Davidson and Smith (2012); and Wolfe (2008).
3   See Salomon (2010) and Taylor (2017).
4   Part of my thesis research for my Master of Arts in Critical Sociology at Brock University.
5   All participant names are pseudonyms.
6   For example, see Davidson and Smith (2012) and Wolfe (2008).

References

Bergenmar, J., Bertilsdotter-Rosqvist, H., and Lönngren, A. S. (2015), "Autism and the Question of the Human," *Literature and Medicine* 33 (1): 202–221.

Davidson, J. and Smith, M. (2012), "Autistic Autobiographies and More-than-Human Emotional Geographies," in D. Spencer, K. Walby, and A. Hunt (eds.), *Emotions Matter: A Relational Approach to Emotions*, 260–279, Toronto: University of Toronto Press.

Fraser, H. and Taylor, N. (2019), "If It Weren't for … Women Having Their Anxiety Soothed through Companion Animal Connections," in L. Gruen and F. Probyn-Rapsey (eds.), *Animaladies*, New York: Bloomsbury.

Gruen, L. (2015), *Entangled Empathy: An Alternative Ethic for Our Relationships with Animals*, Brooklyn: Lantern Books.

Gruen, L. (2018), "Empathy," in L. Gruen (ed.), *Critical Terms for Animal Studies*, Chicago: University of Chicago Press.

Jack, J. (2014), *Autism and Gender: From Refrigerator Mothers to Computer Geeks*, Chicago: University of Illinois Press.

Lion, V. (2016), "Disrupting Temple Grandin: Resisting a 'Humane' Face for Autistic and Animal Oppression" at *Decolonizing Critical Animal Studies, Cripping Critical Animal Studies*, Edmonton, AB, June 21–23.

Salomon, D. (2010), "From Marginal Cases to Linked Oppressions: Reframing the Conflict between the Autistic Pride and Animal Rights Movements," *Journal for Critical Animal Studies* 8 (1/2): 47–72.

Solomon, O. (2015), "'But-He'll Fall!': Children with Autism, Interspecies Intersubjectivity, and the Problem of 'Being Social,'" *Culture, Medicine and Psychiatry* 39: 323–344.

Taylor, S. (2017), *Beasts of Burden: Animal and Disability Liberation*, New York: The New Press.

Wolfe, C. (2008), "Learning from Temple Grandin, or, Animal Studies, Disability Studies, and Who Comes after the Subject," *New Formations* 64: 110–123.

Chapter 7

METAPHORS AND MALADIES: AGAINST PSYCHOLOGIZING SPECIESISM

Guy Scotton

Introduction

Just what sort of a malady is speciesism? For analytic animal ethics, it is a conceptual error—"the unjustified disadvantageous consideration or treatment of those who are not classified as belonging to one or more particular species," to take Oscar Horta's definition (2010, 245)—amenable to logical exposition and rebuttal. However, in societies already structured on speciesist principles, scholars and activists—including, of course, analytic animal ethicists—find themselves in the midst of speciesism as an ideology with a particular cultural history and political economy or, as Weitzenfeld and Joy gloss it, "a complex of institutions, discourses, and affects" (2014, 20). In the struggle to develop a tractable *gestalt* of what I will call "actually existing speciesism"[1] in the midst of the moral emergency of industrial animal exploitation, the animal liberation movement has mobilized a host of overlapping models and metaphors to capture its shifting parameters.

One persistent set of metaphors, with its own ableist connotations, equates moral attention with sight and conceives of speciesism as a wall, barrier, or fence that inhibits or "blinds" the broader public to their "natural" sentiments of concern and respect for nonhuman animals. (For a critical discussion of the "politics of sight" conjured by such metaphors, see Pachirat 2011, chap. 9.) And then there are those metaphors that appropriate concepts of mental illness and disability as pejorative figures for human social relations with other animals.

What I am concerned with here are not claims that particular psychological patterns are reliably associated with speciesist beliefs, attitudes, and actions, but rather claims that speciesism constitutes an overarching disorder, one that shapes both institutions and their participants in ways that are fundamentally pathological, threatening mental and social coherence at one and the same time. Sometimes these associations are concrete and explicit, as when Ingrid Newkirk posits that trophy hunters and serial killers share "the same twisted psychology," pronouncing trophy hunting "the pastime of psychopaths" (2015). In other cases, the associations are more evocative, edging forward the notion that harmfully incoherent or contradictory beliefs are the province of "madness." Judith Butler, reflecting on the paradoxical exceptions generated around the category of "the human" by the interplay between racist and anthropocentric discourses, puts the point suggestively: "This is the kind of thinking that drives people mad, of course, and that seems right" (2015, 35–36).

In this chapter, I argue that this array of metaphors constitutes a diagnostic tendency in animal liberation theory and advocacy—a tendency to construe struggles over medical frames of reference as a battle to be won rather than a paradigm to be challenged in its own right. In their content analysis of major newspapers and activist blogs, Wrenn et al. propose that the medicalization of animal liberation discourse can be understood as a case of "frame contestation," in which social movements and countermovements make their appeals by reconfiguring the culturally available meanings and associations of the issue in question (2015). Because activists and scholars must press their claims in (and otherwise function socially within) terms that are resonant with the dominant culture, it is not surprising to find some advocates struggling for possession, rather than subversion or dissolution, of the authority to call others mad. This "ableism in the ranks" (Wrenn et al. 2015, 1313) has an inward function as well, projecting (if tacitly) regulatory ideals of ability and sanity upon the movement's working definitions of identity and cohesion.

I suggest that engaging critically with this diagnostic tendency in animal liberation theory and advocacy offers a renewed approach to actually existing speciesism, better integrating work in sociology and social psychology with normative ethics and political theory. This critical project also presents an opportunity to rethink the spectrum of abilities within and between species as the grounds for interspecies community. This approach challenges both the premises and aspirations of interspecies justice, opening new possibilities for coalition with disability theory and advocacy.

My chapter is organized as follows. In the first section, I survey some indicative associations made between speciesism and mental pathology in animal rights discourse, ranging from flippant metaphors to analogies and models making substantive psychological claims. Invoking first critiques made by disabilities studies scholars and then the recent emergence of theories of disability from within critical animal studies, I highlight the harms and hazards that this mode of discourse perpetuates. In the second section, I offer two resources that might help animal liberation discourse to move beyond this diagnostic frame. One is a new critical agenda, extending Charles Mills' research program for a naturalized epistemology of (racist) ignorance to speciesism as domination. The other is a proposal for an explicitly interspecies conception of neurodiversity. I conclude the third section with some clarifying remarks and a call for a renewed rhetoric of animal liberation.

Against speciesism as madness

Vivid images can have a long afterlife in academic prose no less than in other forms of literature. In a venerable passage, Ted Benton characterized Marx's depiction of human social relations to nature as "a quite fantastic species-narcissism" (1990, 248). With this phrase, Benton established within animal ethics a trope with a long standing: after all, Freud had considered the Darwinian revolution the second of "two great outrages" against humanity's "naïve self-love," a blow that "rebuked [man] with his descent from the animal kingdom, and his ineradicable animal nature" (1920, 246–247). Benton's phrase continues to echo in the literature as a précis of human–nonhuman relations more generally. Kymlicka and Donaldson, for instance, apply Benton's phrase not to Marx but to the collective "sense of entitlement" and "moral blindness" embodied by societies that condone industrial animal exploitation (Kymlicka and Donaldson 2016, 692).

Gary Francione established another notorious psychological trope within animal ethics, pronouncing the fundamental inconsistencies and contradictions in the benefits and protections human communities offer some animals, versus the harms they systematically inflict on others, a form of "moral schizophrenia" (1996, i). In response to ongoing criticism for his use of the term, Francione has only bolstered his claim: "Some critics argue that it is sufficient to say that our moral views about nonhuman animals are contradictory or confused. No, it's not sufficient.[2] When it comes to nonhuman animals, our views are profoundly delusional and I am using that term literally as indicative of what might be called a social form of schizophrenia" (2009).

In *Rattling the Cage*, Steven Wise builds his case for consciousness in the other great apes in part through a developmental comparison between his (putatively neurotypical) daughter and an autistic child (2000, 152–154). He concludes his book by way of a grand analogy between speciesism and autism, musing: "Perhaps we are an autistic species, biologically incapable of recognizing that nonhumans have minds. Pathologically self-absorbed, we relate to them as if they were machines" (263). Yet for Wise speciesism is an entrenched but permeable moral–religious–juridical belief system, fortified but not determined by humans' general cognitive propensities to interpret other entities and events as serving human purposes: "All this suggests that while our animal mindblindness may be a tendency, even a strong one, it is not irresistible" (264).

Paul Waldau repeatedly employs the trope of autism in a similar way, diagnosing dominant traditions in law, politics, and religion as "almost autistic" (Waldau 2013a, 30), "virtually autistic" (2006, 41; 2010, 81; 2013b, 109), or "truly autistic" (2016, 21; cf. 26) in their regard for nonhuman animals. For Waldau, this "autistic" legacy seems to consist of a particular sort of "self-inflicted ignorance" (2006, 52), involving interpretive distortions not only of the lived realities and individual capacities of nonhuman animals but also of the different social ontologies and modes of relating to other species as communities that are contained within these traditions.

Elisa Aaltola's use of these tropes is instructive, because here they function rhetorically to invert specific forms of philosophical and scientific skepticism about other species' abilities. Thus, for Aaltola, an insistence on mechanical explanations for animal behavior within the personhood debate itself calls the robustness of *human* personhood into question, insofar as personhood is grounded intersubjectively in intercourse with other minds: "Perhaps not recognising animal experiences, and ultimately animal personhood, consists of 'human-autism', which leads to questioning our own capacity for personhood" (2008, 19). The potential social consequences are stark: "Does skepticism [about animal suffering] not push us toward behaving in a psychopathic, narcissistic fashion in our dealings with other animals?" (Aaltola 2013, 463; cf. 2015, 45) Many of the other examples collected here are, in part, tacit or explicit attempts to leverage a similar dynamic, rhetorically 'turning the tables' on those (whether other scholars or society at large) who would deny or diminish the capacities of other animals.

The point of these examples is not to derive a list of proscribed metaphors or modes of thinking about speciesism, but to illustrate a "diagnostic tendency" within animal liberation discourse centered on abject associations with psychological concepts. However, because even the most stringently conceptual language tends toward (and often benefits from or requires) the figurative, and because a full-fledged account of actually existing speciesism must involve some appeal to human moral psychology, I acknowledge that this critique will only be constructive up to a point that is not, in itself, easily delineated. Other theorists, such as Martha Nussbaum and Zipporah Weisberg, offer more thoughtful quasi-psychological models of human exceptionalism that do not depend on specific derogatory analogies, drawing instead from evocative tropes such as denial and repression, and their meanings in psychoanalysis and critical theory. Considering these models at the outset will help to stake out some of the limits of my own diagnosis.

Frans de Waal coined the term "anthropodenial"—"a blindness to the human-like characteristics of animals, or the animal like characteristics of ourselves" (1999, 258)—as a corollary, and corrective, to ethology's perennial worries about anthropomorphism. Nussbaum, extending and modifying de Waal's term, argues that "[h]uman compassion is diseased" (2010, 203), ruptured by a powerful tendency to deny our human creatureliness, and by "the misogyny that is all too often a concomitant of that denial, since women have repeatedly been portrayed as somehow more bodily than men" (206). By "diseased," Nussbaum means not just that human compassion is fatefully limited, subject to partiality and proximity—a limitation that all species share to various degrees. Rather, humans are impelled, via shame and disgust, to profound internal contradictions that manifest in "moral deformity" (203) and, ultimately, pathological violence.

Uniquely deformed, human compassion is in unique need of therapeutic intervention which, as Nussbaum's broader philosophical project details, means cultivating moral attention, including the right moral emotions, to the vulnerability and neediness that unites us as embodied animals. Nussbaum also asserts here that she "mistrust[s] all reductive monocausal accounts of human depravity," but that without further inquiry into anthropodenial as such, "we have little hope of coming up with an adequate account of gendered violence, or of the aspects of violence in general that are implicitly gendered, involving a repudiation of the filth, stickiness, and non-hardness that are the lot of all human beings" (207, 222)—indeed, of all animals.

Weisberg (2011) argues in more expansively critical terms that the psychological dimension of speciesism is best understood as a pathological repression of humans' own animality. Like Nussbaum, Weisberg takes up Freud, but reads him via the early Frankfurt School's account of repression as concomitant with the technical domination of the more-than-human world: "Animal repression can result in or is even constitutive of an unconscious sense of loss, melancholia, ambivalence, guilt, and a host of other neuroses, on both an individual and a societal level" (178). According to Weisberg, this repression manifests not in apathy *simpliciter* but in "neurotic ambivalence": warring senses of love and hatred for animal others, investing our encounters with other animals and their images alternately with yearning and contempt; deep guilt over the incessant enactment of this ambivalence in the mechanized violence of slaughter; and "hysterical indifference" to, and alienation within, systems of industrial animal exploitation (181–183).

This is a specifically patriarchal psychology, in which women (are expected to) embody the uneasy reconciliation of beastly subordination in the category of "domestication" (Weisberg 2011, 183–184). Only through practical reconciliation with animal others, cultivating the relationships and practices of care that are coded as feminine and derided by the patriarchal order of speciesism, can human communities "begin reversing the impact of a repression–oppression complex that is literally suicidal, zoocidal, and ecocidal" (193).

Informed by overlapping critical traditions, both Nussbaum's and Weisberg's depictions of speciesism as psychological malady are resistant to the charge that they denigrate particular psychological categories or disabled identities. Both accounts advance the crucial notion that animal ethics needs a moral psychology that can name speciesism and patriarchy as deeply comorbid conditions. However, insofar as the terms of repression and denial are still liable to construe the logic of disease and contamination in sickness/wellness dichotomies, these accounts too could benefit from explicit attention to critical disability theory in general and to the proposal for a conception of interspecies neurodiversity I make further below.

I return now to the rhetorical stable of schizophrenia, autism, narcissism, and psychopathy. Tenuous and derogatory as many of these psychological tropes may be individually, the first point to observe is that they are all shaped by, and contribute to, complex social histories of abjection and appropriation at which I can only gesture here. A 2003 study of US newspaper articles found that 28 percent of references to schizophrenia were metaphorical, compared to 1.3 percent of references to cancer (Duckworth et al. 2003); similar studies of the metaphorical use of schizophrenia have been conducted in several other countries, with a low of 11 percent in UK newspapers (Chopra and Doody 2007) and a staggering 73.7 percent in the Italian press (Magliano et al. 2011). Illnesses and disabilities are themselves interpreted through metaphors that echo and collide: the concept of "mindblindness," for example, as referenced by Wise (2000) above, originates in and pervades the clinical literature on autism. The term purports to capture the difficulties autistic people face in inferring the mental states of others from standard behavioral cues. However, as Dinishak points out, not only does this metaphor of perceptual deficit further stigmatize blindness as such, it collapses the inherently social context of communicative and interpretive problems: indeed, "non-autistic individuals can also have considerable difficulty understanding the mental lives of autistic individuals" (2013, 75). Waldau's portrayal of autism as a self-consumed, self-serving deficit of human institutions such as economic discourse and public policy recalls the self-described "post-autistic economics" movement that began in France in 2000 as a rebuke to the brittle axioms and shallow models of human behavior espoused by neoclassical economics (Fullbrook 2002).

Disability theorists have criticized extensively the unreflective use of terminology pertaining to mental disabilities and disorders within animal rights discourse, arguing that this indicates a broader disengagement of animal rights theory and activism from the lived experiences and perspectives of those with mental disabilities. With a particular focus on the "argument from marginal cases," these critics have concentrated on the representation of humans with disabilities as providing standards or normative fulcrums by which to extend moral consideration to nonhuman animals, who are taken to meet or surpass the "bar" set by severely cognitively disabled humans as regards morally relevant cognitive faculties (e.g., Salomon 2010; Lewiecki-Wilson 2011).

Another array of critiques—along with new political aspirations and identities—comes from the emerging intersection of critical animal studies with disability studies. Only a few years after Sunaura Taylor "challenge[d] the fields of disability studies and animal rights to take *each other* seriously" (2011, 219; emphasis in original), work in this area has proliferated, with collections offering frames such as "eco-ability" (Nocella II et al. 2017) and "eco-crip theory" (Ray and Sibara 2017), and Taylor's own synthesis of animal and disability liberation (2017). These works demonstrate that when disability is understood as more than a burden or deficit, it can enrich and pluralize cornerstone concepts of political association such as solidarity, interdependence, resistance, and individual and collective agency (Taylor 2017)—insights of profound significance to animal liberation theory and advocacy, if dialogue and cooperation along these lines of inquiry can be sustained.

In light of these developments, could the animal movement nonetheless make strategic and subversive associations between speciesism and madness? After all, as May and Ferri observe, the full force of ableist rhetoric has been wielded "[o]ver the last 200 years" to render feminists and "a wide array of change agents unintelligible, irrational, and unreasonable" (2005, 120). Women's leadership over the past two centuries of animal advocacy has been derided by just such a fusion of misogyny and ableism. Should terms like "madness" and "hysteria" be reclaimed, inverted against an order named specifically as speciesist, ableist, and patriarchal?

The trouble with such inversions, May and Ferri argue, is that they tend to treat the paradigm of disability itself as a fixed point of leverage: wielded as a diagnostic category, "[d]isability is rarely conceptualized as a constructed outcome of power, nor is it regarded as a political identity forged in and through systems of domination" (121). Moreover, abject metaphors evoke complementary images of emancipation: when "metaphors of madness, crippling, and more ... characterize and locate objects of remediation, in this case dominant ideologies, practices, and politics," this often launches a corresponding "use of ableist notions of mobility and movement to define and imagine liberation, resistance, and transformation" (122).

Such a dichotomy between speciesist pathology and vegan praxis[3] threatens to occlude the diversity of moral experiences of those with mental illnesses and disabilities, as well as to construe vegan praxis as a determinate, oppositional state of "wellness" or "sanity." Consider, for example, the dichotomy of dissociation and integration presented by Melanie Joy in her introduction to carnism (2011). Joy posits carnism as the ideological subsystem of speciesism involved in legitimating some animals as objects for human consumption, maintained by an essentially defensive psychological schema.[4] Joy proposes that "[d]issociation is the core defense of carnism, the heart of psychic numbing" (140), and that the forms of "mass dissociation" that sustain carnism are on a psychological "continuum" with the term's clinical manifestations:

> Most of us ... don't dissociate to the degree necessary to kill others; we simply dissociate enough to support the killing that is carried out by others. ... It should come as no surprise that the animals we eat aren't the only ones who pay the price of our dissociation. Dissociation limits our self-awareness and thus presents an obstacle to our personal growth. (141)

Conversely, she concludes, bearing witness to nonhuman animals' suffering—and to our own modes of resistance to such witnessing—"dispels dissociation and leads to a more integrated society" (142).

Distinct from the sheer derogatory association of speciesism with specific psychological and neurodevelopmental traits, the hazard here is that the animal liberation movement, conceived as a movement toward perspectival "integration," will thereby fail to incorporate the perspectives of actual people with (for example) schizophrenia and autism in two senses. First, it may fail to solicit and engage with the distinct perspectives and sensibilities that they might contribute to the critique of speciesism and the praxis of animal liberation; second, it may inadequately account for people's diverse needs for, and ways of, living well within a just interspecies community. Just as vegan advocacy becomes exclusionary—and empirically strained—when it insinuates a necessary or typical harmony between interspecies justice and (some normative conception of) physical health in the form of a vegan diet, so too a notion of emancipatory "wellness," defined in opposition to speciesist pathology, obscures the patent complexities of the mental lives of vegans and carnists alike.

From cognitive and affective domination to
neurodiverse emancipation

Does the case I have made so far against metaphors and models of systemic pathology mean that critical animal scholarship should retreat to a psychologically thin conception of speciesism? On the contrary, I propose that the framing of speciesism as pathological is best complicated through finer-grained sociological and social psychological work on the psychological intricacies of speciesist institutions. Whereas derogatory invocations of madness, schizophrenia, autism, or psychopathy should be renounced by animal liberation discourse, concepts such as ignorance, denial, and repression suggest more sophisticated paradigms that might be approached as interdisciplinary research programs.

One way of framing this agenda could be adapted from Charles Mills' *The Racial Contract* (1997), a work that—like Carole Pateman's *The Sexual Contract* (1988), with which Mills' book is in important conversation—so far has received no concerted attention in critical animal studies. This is somewhat surprising, given the field's commitments to, and extensions of, many of the key insights of feminist and critical race theories. In *The Racial Contract*, Mills conceives white supremacy as a racial contract covertly underwriting institutions of racial domination (contract theory itself, and the mainstream of moral and political philosophy in general, notwithstanding): "On matters related to race, the Racial Contract prescribes for its signatories an inverted epistemology, an epistemology of ignorance, a particular pattern of localized and global cognitive dysfunctions (*which are psychologically and socially functional*), producing the ironic outcome that whites will in general be unable to understand the world they themselves have made" (1997, 18; my emphasis).

While Mills speaks broadly in terms of cognitive dysfunction in a way that could fall back into a metaphor of mental disorder, and indeed is prone to talk of "blindnesses" (19) and "cognitive handicap" (2007, 15), he asserts here that the malady of white supremacy is, in some important sense, psychologically and socially functional. Moreover, this epistemology of ignorance is not just cognitive but somatic and material, as the Racial Contract normalizes and racializes space and the human body (1997, 41–62). This specifically "white moral cognitive dysfunction," Mills suggests, "can potentially be studied by the new research program of cognitive science" (95). Elaborating on this construct in his essay on "white ignorance," Mills proposes a research agenda for a *naturalized* epistemology of ignorance, one that would investigate these cognitive processes of denial and avoidance non-reductively. White supremacy avails itself of what we might regard as forms of extended mind, as Mills elaborates: the simultaneous affirmation of racial norms and their denial as such is scaffolded by social processes such as the public "management of memory" in discourses and artifacts of education, commemoration, and celebration (2007, 28–30).

Grounded in a normative analysis of actually existing speciesism as constituting the systematic oppression of nonhuman animals (e.g., Gruen 2009), it is possible to imagine a complementary epistemology of speciesist ignorance. This project could offer a frame and normative impetus for detailed research into the psychology of speciesist domination that does not rely on problematic analogies to discourses of mental disorder, taking a concerted stance on interspecies justice that disciplines like social psychology and cognitive science typically evade. The resources for such a project have already begun to take shape for critical animal studies. There is now a steady stream of empirical research on human beliefs, moral attitudes, and identities regarding other animals, often organized around the paradigm case of meat consumption (for the most comprehensive review of and prospectus for the field to date, see Amiot and Bastian, 2015). Meanwhile, scholars such as Wicks (2011), Cole and Morgan (2011), and Acampora (2016) have introduced key works in ignorance studies and the sociology of denial to critical animal studies.[5] A naturalized epistemology of speciesism, sociologically thick and non-reductive yet conversant with developments in social psychology and cognitive science, could help to build important conceptual bridges for theories of animal liberation.

Of course, such a framework is still vulnerable to its own ambitions to contribute to a *gestalt* of speciesism, carrying a risk (among others) that it will conceive of "domination" and its epistemology of ignorance as monolithic. One check on this tendency should take the form of an explicit commitment that this research agenda also be reflexive: What patterns of cognition and affect, what flows of ignorance and avoidance shape the animal liberation movement, and how do these interact with other axes of domination? How, for example, does the mainstream of vegan culture propagate white, middle-class "epistemologies of consumption" by promoting dietary "choices" without an adequate account of how access to food and health resources is differentially structured (Harper 2012, 172)? To what extent might characterizations of speciesism as madness and vegan praxis as wellness reflect pervasive cognitive biases and patterns of motivated reasoning of their own, such as "just world" constructs: the expectation, for instance, that wrongdoing is discrete, culpable, and tends to bring about its own punishment?

In criticizing the framing of speciesism as pathology, I have suggested that the emancipatory concepts of "wellness" and "integration" these metaphors evoke in response should also be complicated. The ideal of interspecies justice is in need of concepts and figures that incorporate the experiences and perspectives of people with different psychological profiles into its model of flourishing. One way of framing this contribution would be to develop a conception of *interspecies neurodiversity*. A paradigm forged by the autistic community, neurodiversity signals the diverse goods that a range of differences in mental functioning contributes to, and requires of, the community, if some such range is accepted as a "normal" aspect of human variation rather than a problem to be controlled. This precept illuminates a dimension of justice sensitive to these goods: the prospect of "neuro-equality" (Fenton and Krahn 2009). Neurodiversity has begun to receive concerted attention in ethics and political theory (see, e.g., Herrera and Perry 2013), although many questions and controversies remain regarding the scope and political implications of such diversity.

The idea of interspecies neurodiversity draws a corollary to ideas that have been germinating in the works of neurodiverse scholars for some time: in Salomon's claim, for instance, for recognition of an "autist animal ethics" based on the distinct moral experiences and sensibilities of those with neurodevelopmental differences characteristic of the autism spectrum (2010, 56). Picking up from Salomon, Taylor makes frequent reference to the limits of neurotypical discourse and notes that scholars within neurodiverse communities are now "actively engaging with controversial questions about the relationship between animal and neurodiverse minds" (2017, 114). In this spirit, I want to add a more explicit claim—perhaps too obvious to have merited much discussion—that nonhuman animals are part of the spectrum of neurological diversity. Accordingly, I suggest, the diversity of mental life within and between species should inform aspirations for "neuro-equality," while the concept of neurodiversity should inform conceptions of interspecies justice.

An obvious step through this corollary might now be within the disability/ critical animal studies literature, the notion that morally significant interests and capacities are static, discoverable properties that track species membership in terms of "species-typical functioning" holds sway in much of animal ethics. Thus, Clare Palmer suggests in response that "[nonhuman] animals' capacities have some human relational elements to them. We are not just responding to 'what animals are like' in terms of their capacities; we are actually in part *creating* 'what animals are like'" (2010, 47; emphasis in original). And, I suggest, vice versa: the spectrum of human abilities gains new context and new relational possibilities when arrayed within an interspecies paradigm of neurodiverse ability and difference.

As Sunaura Taylor (2017) details, stigmatic conceptions of disability and animality intertwine to shape what people with disabilities "are like" through scientific, medical, and philosophical discourses that seek to classify and contain disability as deviance from the human prototype. Simultaneously, the institutions and discourses underwriting animal exploitation are both literally and symbolically debilitating for nonhuman animals; a fuller reckoning with "what animals are like," then, must involve understanding nonhuman animals as subjects of, not just objects or figures for, discourses of disability: "Naming animals as crips is a way of challenging us to question our ideas about how bodies move, think, and feel and what makes a body valuable, exploitable, useful, or disposable" (2017, 43). Although Waldau is inclined to portray human abilities as a distinct and coherent profile—we are "creatures of vision embedded in the land" (2006, 44)—he nicely illustrates the interplay between ability understood as a set of individual faculties and the ability to make collective meaning of, and with, those faculties in an interspecies context:

> Crucially, each person's particular heritage of ideas—whatever it is—is no less constraining than the obvious limits conferred on each of us by our limited sensory abilities. Just as we can't hear humpback whales' ever-changing communications in the sea while on a terrestrial path, our forebears didn't tell us about those "songs" because our species had no detailed knowledge of these complex communications' existence in any detail until the last half of the twentieth century. (2006, 44)

This is just the sort of challenge that a critical disability lens sharpens for the moral imagination of animal liberation: "What kinds of experiences and understandings of the world develop for a creature who perceives it through smell or who communicates through bioluminescence? What sort of intelligence is needed to accomplish extremely complex migrations or to survive in the depths of the oceans?" (Taylor 2017, 79). This ongoing project of meaning-making, pushing beyond paradigms of inclusion and accommodation, should inform the conception of flourishing appropriate to a just interspecies society, away from unitary constructs of social and individual "wellness" and toward a neurodiverse politics of living well together.

Toward a renewed rhetoric of animal liberation

I have argued that an inverted critique of speciesism as pathology—and its corollary, tacit or explicit, in understanding vegan praxis as psychic "wellness"—is harmful to humans and other animals with disabilities, impedes solidarity with disability advocacy (and is likely strategically inopportune regarding the broader public), and is internally limiting for animal liberation theory and advocacy. This argument should not be taken to obscure or minimize the fact that speciesist institutions, practices, and attitudes are the systematic *cause* of intense mental distress and illness for both nonhuman animals and humans. I have claimed rather that the social life of speciesism should not itself be conceived on the model of mental illness or disability, precisely because this tends to reinforce a diagnostic frame of disability that occludes the psychological experiences and identities of those—human and nonhuman—who live within regimes of actually existing speciesism.

I have sketched two approaches that might be constructive in moving beyond this diagnostic frame. First, I have argued that resisting this turn to pathology does not involve a retreat from psychological explanation. On the contrary, it affords critical animal studies a richer integration with research agendas in the sociology of ignorance and denial, and in the social psychology of interspecies relations. Such a project need not lose any critical force by dispensing with metaphors of pathology. I have suggested that consolidating the links made to the sociology of denial and ignorance studies, under the auspices of a research program adapted from Charles Mills' account of the epistemology of ignorance enforcing white supremacy, offers one way forward for critical animal studies.

Second, I have suggested that resisting the current of infatuation with the "madness" of speciesism that courses through these metaphors presents an opportunity for a renewed conception of interspecies justice as intricately minded and embodied. By attending to the lived experiences of mental diversity within and between species, critical scholars of speciesism can avail themselves of new theoretical resources and new moral perspectives. In this way, accounts of speciesism might better promote the task of reconstructing moral relations with other animals, in line with the positive commitments staked out by Weisberg and Nussbaum. Informed by the emerging nexus of disability theory and critical animal studies, I have proposed that such accounts of flourishing should incorporate an interspecies conception of neurodiversity.

Finally, although I have focused in this chapter on the ways in which one set of metaphors may denigrate and exclude those with disabilities as well as hamper the analysis of both speciesism and ableism, I do not mean to imply that animal rights theory and advocacy can or should dispense with metaphors or with passionate rhetoric; I agree with Barry Eidlin's declaration that "movements need metaphors" (2013). If, as Michele Moody-Adams argues, conceptual and practical moral progress is necessarily local to some particular domain, gestating within the very cognitive–affective system of situational meanings it aspires to change, then moral growth[6] in some domains might well "require supplying a new metaphor, or some other imaginative structure, in an attempt to reshape our conception of a particular phenomenon" (1999, 175).[7]

The structure of metaphor as such—the pressing together of two apparently unlike things, surprising us with new meaning—itself intimates a figure for neurodiverse interspecies solidarity. Indeed, the conceptual history of solidarity has been shaped by a procession of metaphors—political communities as bodies, families, friends, and teams—whose constraints and ambiguities we may only be able to think through with "the elaboration and introduction of new metaphors" (Honohan 2008, 81). As large as such figures of speech loom in Western political thought, it has often fallen to feminist theory to champion the world-making power of stories, metaphors, and symbols in moral theory, and in the conceptions of political order developed by political theory (see, e.g., Nussbaum 2002, 496–499). One of the tasks that lies ahead for the incipient "political turn" in animal ethics, then, is to return to the critiques of abstract rationality furnished by feminist animal ethicists, and to develop a positive account of how commitments to interspecies justice are shaped and sustained by political rituals, rhetoric, and emotions. Rather than defining itself against specters of speciesism as madness or disorder, a reinvigorated political rhetoric of animal liberation might begin from a sense of the vital and passionate diversity—including neurodiversity—that should be both its ground and its aspiration.

Notes

1   I offer this phrase by analogy to the venerable notions of "actually existing capitalism" and "actually existing socialism," concepts that gesture to the ways in which a society's ideological prescriptions and self-understandings diverge from, and often serve to obscure, the history and functions of its core political–economic institutions.

2   This is despite the fact that Francione begins the essay in which he first employs the term by stating: "Social attitudes about animals are hopelessly confused" (1996, i).

3   Following Weitzenfeld and Joy, "vegan praxis" can be understood here as "a counternarrative and practice in which nonhuman beings are not viewed or treated as appropriate for human consumption" (2014, 21). This involves not just a dietary commitment but the cultivation of morally attentive habits, dispositions, and relationships: "an ever-changing way of understanding and relating to oneself and *all* other beings based on empathy, authenticity, reciprocity, justice, and integrity" (2014, 25). Compare this conception, situated amid the critical animal studies literature cited in Weitzenfeld and Joy's chapter, to Joy's account of "integration" as discussed above.

4   Joy's book, intended for a wider audience, mentions "speciesism" only once in the discussion guide appended to the reprinted edition (2011, 151). Weitzenfeld and Joy situate carnism as "a sub-ideology of speciesism" (2014, 21).

5   In these works, as in much of the related work in sociology and political economy, a different metaphoric regime prevails: as I alluded to in my introduction, talk of visibility/invisibility and blindness/sight abounds. Here, too, scholars should be mindful of the limitations and hazards of their metaphors, even as they examine the particular processes of concealment and revelation that undoubtedly shape our encounters with other animals. Likewise for rhetoric positioning nonhuman animals as "voiceless" and human moral inattention as "deafness," as Sunaura Taylor discusses (2017, 61–67).

6   "Growth" here seems to me a better general figure for moral developments than Moody-Adams' retention of "progress" for a phenomenon that is, on her account, nonlinear and multidimensional. I note, however, that "growth," like "development"

and "progress," also belongs to stultifying metaphoric regimes in economic, business, and policy discourses. These are frames that might be contested via creative links to other biophysical metaphors: for example, those that help to illustrate growth as interdependent maturation (Princen 2010, 64–65).

7   I owe this point to Cooke (2017), who cites Moody-Adams' claim about metaphor in his welcome call for (primarily liberal) animal rights theories to attend in earnest to the moral imagination.

References

Aaltola, E. (2008), "Personhood and Animals," *Environmental Ethics* 30 (2): 175–193.

Aaltola, E. (2013), "Skepticism, Empathy, and Animal Suffering," *Bioethical Inquiry* 10 (4): 457–467.

Aaltola, E. (2015), "Politico-Moral Apathy and Omnivore's *Akrasia*: Views from the Rationalist Tradition," *Politics and Animals* 1 (1): 35–49.

Acampora, R. (2016), "Epistemology of Ignorance and Human Privilege," *Animal Studies Journal* 5 (2): 1–20.

Amiot, C. E. and Bastian B. (2015), "Toward a Psychology of Human–Animal Relations," *Psychological Bulletin* 141 (1): 6–47.

Benton, T. (1990), "Humanism=Speciesism? Marx on Humans and Animals," in S. Sayers and P. Osborne (eds.), *Socialism, Feminism, and Philosophy: A Radical Philosophy Reader*, 237–276, London: Routledge.

Butler, J. (2015), *Notes Toward a Performative Theory of Assembly*, Cambridge, MA: Harvard University Press.

Chopra, A. K. and Doody G. A. (2007), "Schizophrenia, an Illness and a Metaphor: Analysis of the Use of the Term 'Schizophrenia' in the UK National Newspapers," *Journal of the Royal Society of Medicine* 100: 423–426.

Cole, M. and Morgan, K. (2011), "Veganism Contra Speciesism: Beyond Debate," *The Brock Review* 12 (1): 144–163.

Cooke, S. (2017). "Imagined Utopias: Animal Rights and the Moral Imagination," *The Journal of Political Philosophy* 25 (4): e1–e18.

De Waal, F. B. M. (1999), "Anthropomorphism and Anthropodenial: Consistency in Our Thinking about Humans and Other Animals," *Philosophical Topics* 27 (1): 255–280.

Dinishak, J. (2013), "Mindblindness: A Troubling Metaphor?" in C. D. Herrera and A. Perry (eds.), *Ethics and Neurodiversity*, 67–85, Newcastle upon Tyne: Cambridge Scholars Publishing.

Duckworth, K., Halpern, J. H., Schutt, R. K., and Gillespie, C. (2003), "Use of Schizophrenia as a Metaphor in U.S. Newspapers," *Psychiatric Services* 54 (10): 1402–1404.

Eidlin, B. (2013), "The Metaphors of Movements," *Against the Current* 163. Available online: https://www.solidarity-us.org/node/3822 (accessed October 4, 2017).

Fenton, A. and Krahn, T. (2009), "Autism, Neurodiversity and Equality beyond the 'Normal,'" *Journal of Ethics in Mental Health* 2 (2): 1–6.

Francione, G. (1996), "Animals as Property," *Animal Law* 2: i–vi.

Francione, G. (2009), "A Note on Moral Schizophrenia." Available online: http://www. abolitionistapproach.com/a-note-on-moral-schizophrenia/ (accessed October 4, 2017).

Freud, S. (1920), *A General Introduction to Psychoanalysis*, Hall, G. S. (trans.), New York: Boni and Liveright.

Fullbrook, E. (2002), "The Post-Autistic Economics Movement: A Brief History," *The Journal of Australian Political Economy* 50: 14–23.

Gruen, L. (2009), "The Faces of Animal Oppression," in Ferguson, A. and Nagel, M. (eds.), *Dancing with Iris: The Philosophy of Iris Marion Young*, 161–172, Oxford: Oxford University Press.

Harper, A. B. (2012), "Going beyond the Normative White 'Post-Racial' Vegan Epistemology," in P. W. Forson and C. Counihan (eds.), *Taking Food Public: Redefining Foodways in a Changing World*, 155–174, New York: Routledge.

Herrera, C. D. and Perry, A. (eds.) (2013), *Ethics and Neurodiversity*, Newcastle upon Tyne: Cambridge Scholars Publishing.

Honohan, I. (2008). "Metaphors of Solidarity," in T. Carver and J. Pikalo (eds.), *Political Language and Metaphor: Interpreting* and *Changing the World*, 69–82, London: Routledge.

Horta, O. (2010), "What Is Speciesism?" *Journal of Agricultural and Environmental Ethics* 23: 243–266.

Joy, M. (2011), *Why We Love Dogs, Eat Pigs, and Wear Cows: An Introduction to Carnism*, reprint, San Francisco, CA: Conari Press.

Kymlicka, W. and Donaldson, S. (2016), "Locating Animals in Political Philosophy," *Philosophy Compass* 11 (11): 692–701.

Lewiecki-Wilson, C. (2011), "Ableist Rhetorics, Nevertheless: Disability and Animal Rights in the Work of Peter Singer and Martha Nussbaum," *JAC* 31 (1): 71–101.

Magliano, L., Read, J., and Marassi, R. (2011), "Metaphoric and Non-Metaphoric Use of the Term 'Schizophrenia' in Italian Newspapers," *Social Psychiatry and Psychiatric Epidemiology* 46 (10): 1019–1025.

May, V. M. and Ferri, B. A. (2005), "Fixated on Ability: Questioning Ableist Metaphors in Feminist Theories of Resistance," *Prose Studies* 27 (1–2): 120–140.

Mills, C. (1997), *The Racial Contract*, Ithaca: Cornell University Press.

Mills, C. (2007), "White Ignorance," in S. Sullivan and N. Tuana (eds.), *Race and Epistemologies of Ignorance*, 11–38, Albany, NY: State University of New York Press.

Moody-Adams, M. (1999), "The Idea of Moral Progress," *Metaphilosophy* 30 (3): 168–185.

Newkirk, I. (2015), "The Pastime of Psychopaths." Available online: https://www. huffingtonpost.com/ingrid-newkirk/the-pastime-of-psychopath_b_8084410.html (accessed January 19, 2018).

Nocella II, Anthony J., George, Amber E., and Schatz, J. L. (2017), *The Intersectionality of Critical Animal Studies, Disability Studies and the Environment*, London: Lexington Books.

Nussbaum, M. (2002), "Rawls and Feminism," in S. Freeman (ed.), *The Cambridge Companion to Rawls*, 488–520, Cambridge: Cambridge University Press.

Nussbaum, M. (2010), "Compassion: Human and Animal," in N. A. Davis, R. Keshen, and J. McMahan (eds.), *Ethics and Humanity: Themes from the Philosophy of Jonathan Glover*, 202–228, Oxford: Oxford University Press.

Pachirat, T. (2011), *Every Twelve Seconds: Industrialized Slaughter and the Politics of Sight*, New Haven, CT: Yale University Press.

Palmer, C. (2010), *Animal Ethics in Context*, New York: Columbia University Press.

Pateman, C. (1988), *The Sexual Contract*, Stanford, CA: Stanford University Press.

Princen, T. (2010), "Speaking of Sustainability: The Potential of Metaphor," *Sustainability: Science, Practice and Policy* 6 (2): 60–65.

Ray, S. and Sibara, J. (eds.) (2017), *Disability Studies and the Environmental Humanities: Toward an Eco-Crip Theory*, Lincoln, Nebraska: University of Nebraska Press.

Salomon, D. (2010), "From Marginal Cases to Linked Oppressions: Reframing the Conflict between the Autistic Pride and Animal Rights Movements," *Journal of Critical Animal Studies* 8 (1–2): 47–72.

Taylor, S. (2011), "Beasts of Burden: Disability Studies and Animal Rights," *Qui Parle* 19 (2): 191–222.

Taylor, S. (2017), *Beasts of Burden: Animal and Disability Liberation*, New York: The Free Press.

Waldau, P. (2006), "Seeing the Terrain We Walk: Features of the Contemporary Landscape of 'Religion and Animals,'" in P. Waldau and K. Patton (eds.), *A Communion of Subjects: Animals in Religion, Science, and Ethics*, 40–61, New York: Columbia University Press.

Waldau, P. (2010), "Law and Other Animals," in M. DeMello (ed.), *Teaching the Animal: Human–Animal Studies across the Disciplines*, 75–110, New York: Lantern Books.

Waldau, P. (2013a), "Venturing beyond the Tyranny of Small Differences: The Animal Protection Movement, Conservation, and Environmental Education," in M. Bekoff (ed.), *Ignoring Nature No More: The Case for Compassionate Conservation*, 27–44, Chicago: The University of Chicago Press.

Waldau, P. (2013b), *Animal Studies: An Introduction*, Oxford: Oxford University Press.

Waldau, P. (2016), "Second Wave Animal Law and the Arrival of Animal Studies," in D. Cao and S. White (eds.), *Animal Law and Welfare–International Perspectives*, 11–44, New York: Springer.

Weisberg, Z. (2011), "Animal Repression: Speciesism as Pathology," in J. Sanbonmatsu (ed.), *Critical Theory and Animal Liberation*, 177–194, Plymouth: Rowman & Littlefield.

Weitzenfeld, A. and Joy, M. (2014), "An Overview of Anthropocentrism, Humanism, and Speciesism in Critical Animal Theory," in A. J. Nocella II, J. Sorenson, K. Socha, and A. Matsuoka (eds.), *Defining Critical Animal Studies: An Intersectional Social Justice Approach for Liberation*, 3–27, New York: Peter Lang.

Wicks, D. (2011), "Silence and Denial in Everyday Life—The Case of Animal Suffering," *Animals* 1 (1): 186–199.

Wise, S. (2000), *Rattling the Cage: Toward Legal Rights for Animals*, New York: Perseus Books.

Wrenn, C. L., Clark, J., Judge, M., Gilchrist, K. A., Woodlock, D., Dotson, K., Spanos, R., and Wrenn, J. (2015), "The Medicalization of Nonhuman Animal Rights: Frame Contestation and the Exploitation of Disability," *Disability & Society* 30 (9): 1307–1327.

Chapter 8

THE HORRIFIC HISTORY OF COMPARISONS BETWEEN COGNITIVE DISABILITY AND ANIMALITY (AND HOW TO MOVE PAST IT)

Alice Crary

What are we to make of the horrific history of the use of animal comparisons in rhetoric urging the marginalization, abuse, and killing of cognitively disabled human beings? Are there appropriate responses to this history that equip us to insist on the equal moral value of the lives of cognitively disabled human beings without simply reinscribing the denigration of animals in our moral and political discourse? Is it possible to combine an image of animals as in themselves morally significant beings with a commitment to human moral equality? This chapter offers an affirmative answer to this last question, and it proceeds by commenting on a conversation that took place at a 2008 conference, in the Philosophy Department of New York State's Stony Brook University, on challenges that the lives of cognitively disabled human beings pose to widely held philosophical beliefs. The most intense public conversations at the event had to do with what some of the speakers saw as the rebuke cognitive disability represents to the classic idea of the equal dignity of all human life or, more succinctly, to the idea of "basic human equality" (Waldron 2017, 4). These speakers defended views to the effect that severely cognitively disabled human beings merit less solicitude and concern in virtue of their disabilities and, further, that reflection on their lives therefore brings into question traditional egalitarian ideals (see esp. McMahan 2010 and Singer 2010). Some of the conference's other speakers responded with expressions of outrage, protesting that this antiegalitarian stance is not only morally offensive and politically pernicious but also philosophically indefensible (see the account of the conference in Kittay 2010).

A key issue at stake between those who set out to question the idea of human moral equality and those who sought to defend it was the appropriate role of animal comparisons in thinking about the lives of human beings with cognitive disabilities. The original impetus for the—at times very heated—discussion of this particular topic was a lecture given by the high-profile moral philosopher and bioethicist Peter Singer. At Stony Brook, Singer methodically presented a case against the idea of human moral equality, specifically by arguing that some seriously cognitively disabled human beings merit less consideration than their cognitively better endowed human fellows. This is a view that Singer was at the time already well known for championing, and it was already characteristic of his strategy for advocating for the view to make use of comparisons between animals and the cognitively disabled (see, e.g., Singer 1994, 159–163). What distinguished Singer's Stony Brook remarks was, above all, the extent to which he foregrounded such comparisons.

Singer's standard line about the moral standing of severely cognitively disabled human beings rests on the premise that the plain fact of being human is morally indifferent. Singer takes it for granted that no consideration a creature merits can be a direct reflection of membership in a life-form or group like "humans" and that any such consideration must be justified by what some ethical theory enables us to recognize as *grounds* for it. He himself defends a strain of utilitarian theory on which any solicitude that beings warrant is a function of certain individual capacities, in particular, those for pain and pleasure, and he believes that consistency obliges us to allow that any capacities that are morally pertinent in human beings are equally so in animals. *This* is the moment in his thought about cognitive disability at which Singer generally turns to comparisons with animals. He finds it natural at this point to ask whether some humans and some animals are similarly endowed with morally relevant capacities, and he answers his own question in the affirmative. He presents what he sees as evidence for thinking that some human beings who (say, as a result of a congenital condition) are severely cognitively disabled are no better equipped than some animals with "morally significant capacities," and—in a famous flourish—he concludes that a willingness to assign value to the mere fact that a being is human is *speciesist* in the sense of involving unwarranted prejudice in favor of our own species (Singer 2009, 18–23).

At Stony Brook, Singer presented these ideas with particular attention to studies of great apes, dogs, and grey parrots that, he claimed, revealed animals of these types to have cognitive capacities superior to those of human beings with, as he put it in the published version of his conference paper, "profound mental retardation" (Singer 2010, 332–333). Singer is and was an outspoken advocate of animal protectionism (see, e.g., Singer 2009), and one of his aims in appealing to these sorts of animal studies—not only at the conference but quite generally—is to show that some animals merit significantly greater solicitude than they typically receive. Yet, insofar as he addresses questions about ethics and severely cognitively disabled humans, as he did at Stony Brook, his ambition in employing animal comparisons is to show that the cognitively disabled have diminished moral standing and that reflection on their lives speaks against affirming an unqualified claim about human moral equality.

The use of animal comparisons in theorizing about cognitive disability is intensely controversial. Some theorists object to such comparisons even when they are not used, as Singer and a fair number of other contemporary theorists use them (see, e.g., Dombrowski 1997; McMahan 2005, 2008, 2009;

Rachels 1990), to raise questions about the moral standing of cognitively disabled human beings. (For a representative set of the relevant objections, see Drake 2010; Kulick and Rydström 2015, 273; Price 2011, 133; and Spicker 1990.) The objections reflect awareness of a long history of the employment of animal comparisons in rhetoric urging the marginalization, abuse, and killing of these individuals. Discussions of the relevant history often focus on the place of animals in the rhetoric that the Nazis employed in trying to rationalize the wartime mass murder of patients in psychiatric institutions, and it is certainly reasonable to use these atrocities as a reference point. But the employment of denigrating comparisons of cognitively disabled human beings with animals does not begin in Nazi Germany. There is, for instance, evidence that the use of these comparisons was, at various places in late Renaissance Europe, associated with the mistreatment of cognitively disabled human beings (e.g., by housing them in facilities resembling modern zoos and exposing them to extreme temperatures) (Wolfensberger 1972, 17–23). Moreover, even if our main goal is to better understand the logic of the Nazi use of animal analogies, it is helpful to see how some harmful patterns of thought that bring animals together with cognitively disabled human beings, and that wind up informing Nazi propaganda, originate in nineteenth-century European conversations.

The most notable of these patterns are connected to lines of reasoning that receive a decisive formulation in Darwin's *Descent of Man* ([1871] 1909). Darwin repeatedly likens cognitively disabled human beings to animals, and he does so with an eye to accounting for the "missing link" between human beings and their closest evolutionary antecedents (see Gelb 2008). Although Darwin cannot provide evidence of whole life-forms between "civilized" human beings and apes, he thinks he can demonstrate kinship between what he regards as the most sophisticated nonhuman animals and the least developed human beings (see, e.g., Darwin [1871] 1909, 98–100; for Darwin's general concern with "the imperfection of the geological record" in confronting us with an "absence of intermediate varieties," see Darwin 2006, esp. chapter 9). Darwin's outlook here is monogenistic, positing a common origin for all human beings, and there is good evidence that he defended monogeny because he abhorred slavery and wanted to resist the polygenistic defenses of it that were common at the time (see 2009, esp., 24 and 1008 and 1009). Yet he gives his overarching project an explicitly racist inflection insofar as he favors the idea of a linear human scale on which some "races," though members of one human family, are yet "lower" and more "savage" than others, and insofar as he picks out as the "lowest" or most "savage" human cases groups of human beings, identified as "nonwhite," such as "the negro and the Australian [aboriginal]" (Darwin [1871] 1909, 241–242; see also 40–43, 141–142, 156, 169–171 and 180–183; see also Desmond and Moore 2009, 977–982). At the same time, echoing the views of prominent contemporaneous anthropologists (see Gelb 2008), he situates the human beings he calls "idiots" or "imbeciles" outside the human family altogether, depicting them as throwbacks to stages on the evolutionary path to human beings (Darwin [1871] 1909, 21–23 and 53). For Darwin "idiots" are most closely related to nonhuman animals (Darwin [1871] 1909, 53, 102–103, and 131–132) and are as such evidence of losses internal to the workings of natural selection.

There is a nuance to Darwin's own views about the appropriate treatment of "idiots." Darwin holds that we have an indirect duty to care for these individuals, since, as he sees it, to do otherwise would risk "deterioration in the noblest part of our nature" (Darwin [1871] 1909, 205–206). Yet he also warns that allowing "idiots," or their "nonwhite" "savage" human cousins, to "propagate their kind" will lead to "degeneration" of the human species (Darwin [1871] 1909, 206). So, there is a sense in which he himself sets the stage for some of the racist and ableist eugenic fervor of late nineteenth and early twentieth-century Europe (see also Darwin [1859] 2006, 208–209 and 217–220).

One characteristic expression of this fervor can be found in the writings of Ernst Haeckel, a German zoologist and popularizer of Darwin who gives his decidedly racist strain of Darwinianism a recognizable "social" turn insofar as he champions, with regard to the human case, "the idea that one [can] steer the process of natural selection" (Burleigh 1994, 13; for Haeckel's calls for such "steering," see, e.g., Haeckel 1914, chapters 7 and 8). Many of Haeckel's colleagues interpret this "steering" to include the sterilization and isolation of the cognitively disabled, and Haeckel himself endorses killing in some cases (Burleigh 1994, 13). Although there are many conditions contributing to a cultural environment in which it seems reasonable to thinkers like Haeckel to entertain policies for the murder of their cognitively disabled fellows—including the post–First World War rise of "an economically driven quest for enhanced efficiency and rationalization" (Burleigh 1994, 33)—Darwinian tropes about these individuals as subhuman animals play a significant role. Nor is it only in Germany and other European countries that, in the decades before the Second World War, social Darwinist conversations linking cognitive disability to subhuman animality add to the social vulnerability of the cognitively disabled. In the United States, a notable inheritor of late nineteenth-century European eugenic ideas, there were new programs for "managing natural selection" through the mass sterilization of people with cognitive disabilities and the imposing of marriage prohibitions, as well as through immigration restrictions and targeted deportations (see Baynton 2001, 45–46, Gallagher 1995, 84 and 78–80, Gill 2015, 14, and Simplican 2015, 58). Like their European counterparts, these calls for control of individuals with cognitive disabilities seamlessly integrate animal comparisons with racist ideas of hierarchies of human groups (Samuels 2014, 176–178). This is also true in Britain where in the 1880s the physician John Langdon Down classified individuals with what we now—since the 1960s—call Down syndrome as "Mongoloid idiots," depicting them as recursions to what he regarded as an "inferior" race (Baynton 2001, 36, and Simplican 2015, 58 and 63).

Granted these ideological tendencies, we can represent what was done to those with cognitive disabilities in Nazi Germany as an "application" of already widely accepted "principles of Social Darwinism and the budding science of eugenics" (Gallagher 1995, 5). The most horrific application is the wartime program "T4"—named after the Berlin address of its administrative home (*Tiergartenstraße 4*)—under the rubric of which 200,000 people were killed between 1939 and 1945. In the years running up to this period, German asylums were deprived of resources to such an extent that their inhabitants often struck visitors like animals or "beasts" confined in squalid cages (Burleigh 1994, 44), and in memoranda circulated during the planning of T4, prospective grounds for killing "people suffering from serious congenital mental or physical 'malformation' included that they are 'situated at the 'lowest' animal level' (Burleigh 1994, 98) and that to expend resources on them would therefore be to ['sin'] against the 'law of natural selection'" (Burleigh 1994, 195). The Nazis also publicly used animal tropes to sell the murderous T4 policies, for instance, in a series of propaganda films made between 1936 and 1941. Here individuals with intellectual disabilities or chronic psychiatric conditions were often, in the words of historian Michael Burleigh, "explicitly situated below the level occupied by animals, who are invariably depicted with greater affection and sensitivity" (Burleigh 1994, 194). In the 1936 film *Hereditarily Ill [Erbkrank]*, "a shaven-headed youth is shown eating handfuls of grass" (Burleigh 1994, 194), and in the 1941 *I Accuse [Ich Klage an]* a crude social Darwinism is deployed to clear the way for the idea that the involuntary euthanasia of "useless" members of society is as innocuous as putting down a suffering pet (Burleigh 1994, 206–207). A second version of *I Accuse*, which adopts the same basic strategy, includes this bit of dialogue: "Gentlemen, when we foresters have shot an animal and it continues to be in pain, then we give it the *coup de grâce*" (Burleigh 1994, 201; for a remark on this analogy to putting down pets, see also Gallagher 1995, 18). Moreover, in developing their animal-themed and fundamentally social Darwinist approach to T4, the Nazis increasingly lumped race and various forms of perceived social and sexual deviancy together with cognitive disability, treating all as markers of lack of social "fitness" (see, e.g., Gallagher 1995, 77).

The kinds of animal-cognitive disability comparisons that function in nineteenth-century eugenic and social Darwinist thought, as well as in its later Nazi versions, have a distinctive logic. They don't simply involve, on the one hand, associations of cognitively disabled human beings with a specific kind of animal and, on the other, associations of animals of that kind with some significantly negative characteristic. Cognitively disabled human beings are sometimes denigrated by association with traits that are precisely valued in the animals who possess them. We find a similar structure in animal comparisons the Nazis used to denigrate Jews (see, e.g., Raffles 2007, 525). Thus, as one scholar puts it, "the Nazis were deeply attached to their dogs ... [but] that did not keep them from calling Jews 'Hunde'" (Kittay 2010, 399; see also

Gallagher 1995, 255–260, and Kittay 2005, 125, for remarks on Nazi laws for the protection of animals). We can understand the apparent paradox here by referring it back to aspects of Darwin's reasoning. The author of *The Descent of Man* is happy to dehumanize "idiots"—and the racialized "savages" he takes as their close kin—for resembling animals like monkeys whom he praises for their human-like qualities (Darwin 2006, 404–405). This stance makes sense if we take the hierarchy of life-forms internal to evolutionary theory, and also the role of variation in creating it, to be endowed with normative significance (see, e.g., Gelb 2008). Now individual animals of "lower" kinds can be seen as exemplary and laudatory individuals insofar as they possess particular human-like traits that, as it were, "raise" them above their station, while at the same time individual cognitively disabled humans who are taken to possess similar traits—traits that seem to "lower" their station—can be looked down on as evidence of the waste of natural selection.

While the second half of the twentieth century witnessed a large-scale cultural abandonment of social Darwinist beliefs, it's not implausible to think that some of the eugenic practices once associated with social Darwinism (e.g., the forcible control of the sexuality of the cognitively disabled [see Gill 2015], the directing of social resources toward what are perceived as "worthier lives" [see Bodey 2017], and the allocation of medical resources in particular [see Kittay 1999, 164–165]) have outlived these beliefs. Setting aside an investigation into whether, or to what extent, the eugenicist ideas internal to classic comparisons between animals and the cognitively disabled are enjoying a real-world afterlife, it remains important to find ways—not haunted by these ideas—for bringing cognitively disabled human beings into moral thought.[1] Indeed, the appropriate standard for moral thought about the cognitively disabled is higher than this. Such thought should be not only undistorted by these and other forms of prejudice but also informed by the sorts of faithful images of the worldly lives of cognitively disabled human beings that are relevant to ethics.

Singer expresses confidence that his own thinking about cognitive disability meets this high standard. It is not merely that he presents himself as free of pernicious prejudices, such as the "racist viewpoint" that led the Nazis to the "murder of people considered unworthy of living" (Singer 1991, 7; see also Singer 1992, 86). Singer is also sanguine that he is working with accurate empirical descriptions of aspects of the lives of the cognitively disabled. His assurance here is in large part a function of his preferred image of how the mind makes contact with the world. Singer helps himself to a philosophical account of the demands of getting the world in focus—an account that, in its basic outlines, is enormously influential within Anglo-American analytic circles—on which our culturally and ethically local perspectives have an essential tendency to interfere with our view of how things stand, and on which we hence approach a less distorted vision of things by progressively sloughing these perspectives off. Bearing in mind that this account represents movement toward ethical neutrality as essentially tending in the direction of a less obstructed view of the world, we might refer to it as a *neutral conception of reason* (for a detailed discussion of the relevant rendering of world-directed thought, see Crary forthcoming[a]). Insofar as this neutral conception suggests that we approach a more accurate image of empirical reality by stepping back from our ethical attitudes, it implies that nothing in the empirical world can, qua observable, merit such attitudes and accordingly that nothing in the empirical world can, qua observable, be ethically significant.[2]

Singer only occasionally flags his reliance on the neutral conception, but it structures his thought throughout his career. This is true even though his core philosophical contentions have undergone a notable change over the past twenty years. Whereas early on Singer advocates a type of ethical non-cognitivism (i.e., an outlook on which there are no moral properties and on which moral statements aren't exercises of predicating properties) in reference to which he straightforwardly rejects the idea of objective ethical values, more recently he has shifted toward favoring a rational intuitionist stance that qualifies as a form of cognitivism (or objectivism) about values insofar as it allows that some objective "ethical" truths reveal themselves to intuition (De Lazari-Radek and Singer 2014; see also Singer's own commentary on his development in Singer 1999a and 1999b). Singer now allows that the world contains ethical truths in the form of normatively loaded precepts that speak for particular modes of conduct. This does not, however, represent a retreat from the neutral conception of reason. Singer is not suggesting that the precepts he takes to be intuitable only show up in terms of, or are internally related to, specific attitudes. He construes our mental access to these precepts in neutral terms, and, as a result, he regards our grasp of them as lacking the motivational significance that would by itself explain our acting (De Lazari-Radek and Singer 2014, 197–199). So, Singer continues to abide by the constraints of the neutral conception of reason. Moreover, since the only objective "ethical" truths he recognizes are practical precepts, and since, unlike ethical values as they are standardly conceived, these truths are motivationally inert, he effectively retains the view that nothing in the empirical world is, qua observable, in itself ethically significant (for a fuller treatment of these aspects of Singer's work, see Crary 2016, 19–25).

These philosophical commitments are what underlie Singer's claim that the plain fact of being human, taken (as he in fact takes it) as an observable or theoretically available circumstance, is morally indifferent. At the same time, they are part of the motivation for his insistence that, if we are to represent human beings as in themselves meriting respect and attention, we need to find theory-leveraged grounds for human moral status (e.g., according to the kind of utilitarianism Singer favors, individual human beings' capacities for pleasure and suffering). Further, the neutral conception is the source of his faith in the empirical accuracy of things he says about the cognitively disabled. Because he relies on such a conception, he thinks he goes a long way toward ensuring that he is doing justice to any features of the world that interest him in ethics—including the various features of the worldly lives of cognitively disabled human beings that he discusses—simply by thinking and talking about his subject matter in a manner free from noxious forms of prejudice. *This* is why Singer thinks his use of comparisons between human beings with cognitive disabilities and animals is unobjectionable. Since the terms in the comparisons are capacities possessed by individual human beings and animals, and since Singer believes that his preferred ethically neutral methods suffice for bringing such capacities, together with likenesses among them, into focus in ethics, it appears to him that his use of the comparisons is philosophically unimpeachable. To be sure, there is a further pragmatic question about whether in thus effectively dealing in analogues of the sorts of animal comparisons that were once integrated into murderously ableist and racist Nazi policies, Singer—despite his personal abhorrence for the policies—risks contributing to a political climate more welcoming of them. But Singer is relatively dismissive of the idea that his views "could erode respect for human life and so lead to a return of the mentality that made Nazi atrocities possible" (Singer 1990, 42; for a similar observation about Singer's attitude toward the risks here, see Kittay 2017, 32). He takes himself to be justified in concluding that his use of animal comparisons is morally unproblematic as well as philosophically sound.

Is Singer's favorable assessment of his own animal-indexed theorizing about cognitively disabled human beings too charitable? The assessment depends for its apparent plausibility on the neutral conception of reason with which Singer operates, and some of Singer's most outspoken critics attack this conception (see, e.g., Crary 2016, chapter 4; Diamond 1978 and 1991; and Mulhall 2009, chapter 9). Indeed, the credentials of the conception were what was at stake in a striking exchange that Singer had with Eva Feder Kittay, a prominent moral philosopher, feminist theorist, and advocate for the cognitively disabled, during the question-and-answer session after Singer's talk at the 2008 Stony Brook conference. Kittay, one of the conference organizers, was at the time already well known as a thinker who became interested in issues affecting the cognitively disabled while advocating for social support and recognition for her daughter Sesha, who has cerebral palsy and who, in adulthood, doesn't speak and isn't capable of even minimal self-care. Writing after the conference, Kittay declares that "for a mother of a severely cognitively impaired child" it is "devastating" to "read texts in which one's child is compared, in all seriousness and with philosophical authority, to a dog, pig, rat, and, most flatteringly, a chimp" (Kittay 2010, 397). Kittay attributes Singer's willingness to deal in animal comparisons in the way he does to limitations in his grasp of the empirical lives of cognitively disabled human individuals like Sesha. But her philosophical image of what it is to do justice to the empirical world is very different from his.

Kittay inherits from and participates in traditions of feminist theorizing that resist the idea—distinctive of the neutral conception—that neutrality is a regulative ideal for all world-directed thought. Antipathy to this idea is sometimes associated with skepticism about objectivity, and it is certainly true that many post-structuralists, including some who are self-identified feminists, move seamlessly from repudiating an aspiration to neutral modes of criticism to embracing such skepticism. But this post-structuralist stance threatens to be politically disempowering, stripping us of any license to claim genuine authority for our critical claims (see Lovibond 1989). There is also a sense in which the stance rehearses the logic of the very neutral conception from which it supposed to free us, and in which it is thus philosophically less radical than it may at first appear. In taking the forfeiture of the ideal of neutrality as equivalent to the loss of an unqualified claim to objectivity, it effectively appeals to the very neutral conception of reason that it is alleged to be disavowing, albeit while also treating the ideal of neutral mental contact with the world internal to the conception as forever beyond reach (see Crary 2018).

Kittay herself works with notable strands of second- and post-second-wave feminist thought that adopt a more antagonistic attitude toward the neutral conception. Here the conception is taken to be philosophically bankrupt and hence incapable of leveraging an attack on the cognitive credentials of particular modes of thought simply because they are ethically non-neutral. The upshot is that it appears wrong to insist that the features of the empirical world that are relevant to this or that bit of feminist or other ethical reflection must reveal themselves to an ethically neutral gaze (for discussion of the relevant feminist literature, see Crary 2001, 2002, and Crary 2018). This is the basic philosophical outlook that Kittay is presupposing in talking about shortcomings in Singer's understanding of the empirical lives of cognitively disabled human beings like her daughter. She is presupposing that specific modes of affective response—of sorts that might be inculcated, for example, by actual experience interacting with cognitively disabled people—are necessary prerequisites for bringing these aspects into focus (Kittay 2010, 406–407; see also Kittay 1999, chapter 2).

At the time of the Stony Brook conference, Kittay was aware that the Princeton, NJ-based Singer had brought some of his students to a local neonatal intensive care unit in New Brunswick, New Jersey, to look at cognitively disabled newborns. She proposed that the next time he take students on a field trip, he bring them to visit the community of small group homes in which Sesha lives. When Singer demurred, saying that this would be "a little further than New Brunswick," Kittay replied that "she would be happy to personally arrange it" and that she wanted him to "see some of these people that [he was] talking about" (quoted in Kittay 2010, 405). Singer then issued the following challenge to Kittay. "I would like you to tell me—just in terms of the argument that I presented—what it is that I would see there that would challenge the argument" (Kittay 2010, 405).

Singer's challenge was in part a demand that Kittay supply him with an advance on the kind of illumination that her proposed excursion would bring with it. This demand seems reasonable if we insist, in accordance with the neutral conception of reason he himself accepts, that any new aspects of the lives of Sesha and her friends must be accessible to us in a manner that doesn't depend on the sorts of capacities for emotional responsiveness that we might develop, say, by spending time with them. Further, insofar as the demand thus appears justified, the reply Kittay gave Singer is bound to seem quite inadequate. What she said to Singer was in essence that "how much you see is ... what you bring to the situation" and that therefore she was not "sure what [he'd] see" (Kittay 2010, 405). This struck Singer as mere evasive hand waving, and, when his public debate with her continued in the conference's last session, he told her peremptorily that she had to "put up or stop saying that [he doesn't have the empirical stuff right]" (Kittay 2010, 408). Singer's exasperation reflected his thought that he was justified in insisting not only that she tell him what he would see if he visited her daughter's home but also that she do so "in terms of the argument" he presented. This further requirement also seems reasonable if we impose on ourselves the constraints of the neutral conception of reason. Granted these constraints, it seems reasonable to accept two of the main premises of Singer's argument, namely, that the sheer fact of being human (insofar as that is considered to be an observable circumstance) is not in itself morally significant and that we therefore need to find grounds for human moral status.[3] Now it seems fair to demand that Kittay specify not only what Singer will see if he visits Sesha but also how *that*—whatever it is—serves to ground a claim about Sesha's moral standing.

But the logical dynamic of the conversation between Singer and Kittay shifts if, in accordance with the feminist traditions in which Kittay is working, we reject the neutral conception of reason. To be sure, it matters what we identify as an alternative to this conception. Any successful case for an alternative needs to start from an attack on the view that undistorted thought about the world is beholden to an ideal of ethical neutrality, and it is possible to mount a plausible challenge of this sort (see Crary 2016, section 2.1, and Crary forthcoming[a]). But, once we establish that empirical thinking is not as such regulated by an ideal of neutrality, we still need to determine which aspects of the world are such that we need ethical resources to get them in view. Within classic Anglo-American philosophy of the social sciences, there are arguments for thinking that *social phenomena*, taken to be composed of actions that are themselves understood as expressions of practical rationality, are not as such available to thought except through the lens of ethical considerations (see, e.g., Winch 1990, esp. 98–99; for a detailed commentary on relevant portions of the Winch, see Crary 2018e). Such arguments do not, however, suffice for Kittay's purposes. For what distinguishes her position is the thought that ethical resources are required to bring into focus not only the conduct of rational individuals but also aspects of the lives of humans—like Sesha—who never reach rational maturity.

Although Kittay herself doesn't pursue this line of inquiry, it is possible to find considerations that are adequate for a defense of this thought in a view of mind that is sometimes associated with the later philosophy of Wittgenstein. At issue is a view on which our categories for thinking about *all* aspects of mind, whether or not they are expressive of rationality, are ethically inflected categories that resist meaningful translation to neutral terms, and on which these categories are also essentially matters of sensitivity to how things empirically are. Ian Hacking comes very close to making a case for this sort of Wittgensteinian view, arguing—in a series of articles on autism—that the practical adjustments and responses we typically acquire in the course of early socialization contribute internally to our ability to non-inferentially perceive the psychological significance of the real expressions and conduct of others, without regard to whether they have humanly typical or atypical mental capacities (Hacking 2009a, b and 2010). It is possible to elaborate the sort of approach to mind Hacking defends so that it becomes one on which we require the relevant sorts of practical adjustments and responses in order to pick out the psychological character of the expressive behavior of those adult human beings who are atypical specifically in that they to a greater or lesser degree lack mature capacities of reason (see Crary 2016, sections 2.1–2.3). There is, moreover, an additional element of Hacking's work that deserves mention here. Drawing on his preferred Wittgensteinian view of mind, Hacking advocates what he regards as the further Wittgensteinian idea that the modes of practical response internal to psychological understanding encode a sense of what is humanly important. His thought is not merely that we need ethical resources to do justice in ethics to typical or atypical psychological aspects of human existence but also that in order to get these aspects empirically into focus, we need to look at individuals through the lens of an ethical conception of human life. Once again, it is possible to extend Hacking's reasoning, showing that this thought of his (namely, that we need a gaze saturated by a sense of what matters in human life in order to do justice to the psychological expressions of typical or atypical others) holds in connection with our efforts to pick out aspects of the minds of others who are atypical in the specific sense of being nonrational (see Crary 2016, section 4.2).

Suppose we bring this slightly extended version of Hacking's Wittgensteinian reflections to bear on the case of Kittay's daughter Sesha. Part of the idea would be that in order to capture Sesha's worldly circumstances in a manner relevant for ethics, we need to see her in the light of a morally charged image of human existence. Relatedly, we cannot exclude the possibility that there are aspects of Sesha's circumstances that we will fail to register unless we further develop our current ethical image. There is an open-ended number of ways in which we might go about refining this image: by interacting with people with or without cognitive disabilities or, alternately, by engaging with literature, memoirs, journalism, films, or other works of art that shape our attitudes about the lives of people with cognitive disabilities or about human life more generally (see Kittay 2010, 408). We might turn, for instance, to work that the journalist Katherine Boo did for the *Washington Post* in the 1990s, exposing fatal forms of abuse and medical neglect of cognitively disabled individuals in a network of private homes in the D.C. area (Boo 1999). Part of what gives Boo's work interest here is that she takes her commitment to writing about urgent social and political matters like these to be consistent with "experimenting with form" in order to engage readers and expand their conceptions of human importance (this last reference is to remarks of Boo's at the Princeton University Center for Human Values, November 8, 2017). Not that every bit of writing or other oeuvre that aims to enrich our sense of what matters in human life will, insofar as it grabs us, directly inform our ability to do justice in ethics to the worldly existence of others. Any particular work may turn out to be sensationalistic, sentimental, or distorting in some further way. The point of—the extended version of—Hacking's Wittgensteinian reflections is simply that the engaged task of investigating different conceptions of human flourishing is internal to world-directed thought about human beings in ethics without regard to the nature or level of their cognitive abilities. So, we must be working with such conceptions if we are to be in a position to recognize whether, say, a given expression of Sesha's is one of anxiety or affection, whether a given activity with her is exploitative or engaging, or whether a given institutional setting for her represents enrichment or an abusive form of confinement. This is the backdrop against which Kittay invited Singer on an excursion on which he would interact with Sesha and her friends and—as Kittay hoped—thereby develop new modes of responsiveness to light up otherwise inaccessible aspects of their lives.

Now it should be clear that Kittay was right not to accede to Singer's request to confine her response to him to the terms of his argument. What emerges from the line of thought just traced out is that it is only from perspectives afforded by an ethically charged conception of human life that it is possible to make sense of psychological aspects of the worldly existence of human beings, whatever the level or nature of their individual cognitive capacities. Seeing the expressions of a human being is, according to this line of thought, inseparable from seeing the person in question as an individual for whom, simply in virtue of her humanity, certain kinds of things matter. This means that human beings, however well or poorly endowed cognitively, figure in legitimate, world-guided moral thought as beings who merit specific forms of attention just as the kinds of beings they are. So, there is no reason to follow Singer in taking as an argumentative premise the idea that the plain fact of being human is morally unimportant. Additionally, there is no reason to follow him in searching for grounds that some ethical theory enables us to recognize as endowing human beings with moral standing. And, lastly, since there is no reason to search for such grounds, there is a fortiori no reason to accept the terms of an argument, like Singer's, that urges us to look for such grounds in individual human capacities that seem to invite comparison with the individual capacities of other animals.

This last reflection is worth underlining. It's not merely that there is no room in responsible moral thought about human beings with cognitive disabilities for Singer-style animal comparisons. The entire Singerian exercise of appealing to animal comparisons—and a fair number of thinkers have followed Singer in attempting this questionable exercise (see, e.g., Dombrowski 1997; McMahan

2005, 2006 and 2009; and Rachels 1990)—is not so much a contribution to the search for the value in human life as a reflection of the failure to see this value where it is.

Granted that human beings enter moral thought as meriting respect and solicitude just as the sorts of being they are, does it follow that we are obliged to regard animals as "lesser" creatures and, by the same token, to reject *all* human-animal comparisons in ethics as necessarily insidious? Here it is helpful to recall that a key element of the account of the value of bare humanity just presented is a philosophically unorthodox—Wittgensteinian-Hackingian—view of mind. At the heart of this view is the idea that our concepts for aspects of mind, irrespective of whether these are expressive of rationality, are irredeemably ethical and, at the same time, essentially revelatory of how things are. A good case can be made for elaborating a version of this view not only in reference to human minds but also in reference to animal minds (see Crary 2016, chapter 3, and Gaita 2002). Hacking suggests, as we saw, that an exploration of the view in connection with human beings reveals that if we are to do empirical justice in ethics to psychological features of human life, we need to look at individual human beings in the light of an image of what matters in human life. It would be possible to show that an investigation of the view in connection with animals yields an analogous result. The idea would be that if we are to do empirical justice in ethics to the minds of animals of a particular kind, we need to look at individual animals of that kind in the light of an image of what matters in their lives. This would mean that seeing the expressions of an animal of a given kind is inseparable from seeing the creature at issue as an individual for whom, simply in virtue of being an animal of that kind, certain sorts of things matter. It would mean that animals of different kinds enter into sound moral reflections as beings who call for specific forms of attention not as "lesser" beings but just as the kinds of beings they are.

This view of the moral standing of animals, like its counterpart view of human moral standing, has significant implications for how we conceive the demands of moral thought about animals. It implies that investigating different conceptions of flourishing for animals of a given kind is internal to world-directed thought about animals of that kind (see Crary 2016, section 4.3, and Crary 2018). Just as we may move toward a more just understanding of the worldly lives of human beings with cognitive disabilities by immersing ourselves in memoirs, journalism, literature, or other works of art that shape our sense of what matters in human life, we may move toward a more just understanding of the worldly lives of animals of different kinds by immersing ourselves in work that shapes our sense of what matters in the lives of animals of particular kinds. We may progress toward a better understanding here, say, by turning to some of Tolstoy's nonfictional writings on vegetarianism or to Jonathan Safran Foer's work on eating animals. What makes the work of these particular authors especially salient here is that they self-consciously go beyond presenting plain facts of slaughter and try to engage us in ways that will enable us to appreciate the momentousness of the killing of animals (see Crary 2016, sections 6.2 and 7.1, and Crary forthcoming[b]).

Setting aside these methodological issues for other occasions, it is worth noting that the introduction of the above account of the moral standing of animals of different kinds eliminates a hierarchy or normative ordering of life-forms of the sort that informs social Darwinist thought. Now animals of particular kinds show up for us as mattering just as the kinds of creatures they are. So—as far as ethics is concerned—there is no reason to speak, in normatively drenched, pseudo-evolutionary terms, of "lower" and "higher" animals. The point is not merely that we should jettison the type of hierarchy of life-forms found in social Darwinist enterprises. There is an unacceptable hierarchizing of life-forms in the medieval *scala naturae*, and there is a likewise untenable hierarchizing of life-forms in Singer's thought. Despite his hostility to speciesism, Singer presents us with a human-headed ranking of life-forms insofar as he ties creatures' moral status to their individual capacities. For it is platitudinous that those capacities will typically be those that are typical for whatever species is in question, with members of the human species typically receiving the biggest "morally relevant" portion (see Wolfe 2008, 118, for a version of this same idea about Singer's project). The larger point is that there is no room for any normative pecking order, whether a medieval one, a social Darwinist one, a Singerian one, or one of some other kind.

It would be wrong to protest that we *need* an ethically laden hierarchy of life-forms to defend the idea of basic human equality. It is true that classic strategies for exalting humanity involve assigning us a station above that of animals (see, e.g., Maritain [1944] 2012, 66 and 101–2), and, despite the intensity of attacks over the last half century on the sorts of hierarchy of life-forms presupposed by these strategies, it is also true that it remains common today for thinkers to attempt to demonstrate the value of humanity by suggesting that all human beings are superior to animals (e.g., Kateb 2011, 3–4 and 22–24, and Anderson 2014, 494–496). The persistence of this trend is perhaps unsurprising given that many social justice movements—including, among others, the US disability rights movement (for commentary, see Taylor 2017, chapter 2)—demand respect for the oppressed by insisting that they are not animals. It is, however, wrong to think that we can't combat ableism or other forms of oppression without trampling on animals (see, e.g., Taylor 2016, chapter 2). It's not merely that, as the line of thought developed here shows, we can represent all human beings as in themselves valuable without denigrating animals. More significantly, we undermine our own human-centered liberating efforts if we don't do this. Without even following up on empirical evidence indicating that "the more sharply people distinguish between humans and animals, the more likely they are to dehumanize other humans" (Kymlicka 2017, 13), we can say that, in a cultural context like ours, in which the belief that human beings are the end products of a process of natural selection is widespread, the idea of a normative ordering of life-forms is hazardously likely to get associated with the idea of evolutionary stages and, by the same token, hazardously likely to get associated with the idea that some individual human beings or groups of human beings are "lower" than others.

If we abandon a normative ordering of life-forms, we open the door for non-demeaning comparisons between aspects of human life, on the one hand, and aspects of animal life, on the other. Admittedly, Singer's writings about these matters notwithstanding, there are very good pragmatic reasons to be cautious in the use of such comparisons. Given that the notion of ethically charged distinctions between "lower" and "higher" animals is in fact still rampant in our culture, and given that this notion has historically been, and still is, integral to rhetoric that contributes to the subjugation of socially vulnerable groups of people—including, not only the cognitively disabled but also the physically disabled, people identified as nonwhite, women, the gender-non-conforming, the very old, etc.—we need to be especially careful in tracing out lines of filiation between the lives of animals and the lives of members of human groups who confront systematic forms of bias. But we can say all of this without implying that animal likenesses have no place in ethical thought. Indeed, we can say it all without implying that these likenesses have no place in ethical thought that touches on the lives of cognitively disabled human beings.

This is a possibility that Kittay seems to overlook, though, admittedly, her views about the moral standing of animals aren't well developed. In one recent essay, she does say that she sympathizes with challenges to "the claim of [human] superiority [over animals] and its concomitant right to dominate [them]" (Kittay 2017, 31). But elsewhere she seems to at least implicitly operate with the sort of ethically saturated notion of "higher" and "lower" life-forms that would render animal comparisons not merely contingently but inherently demeaning (see, e.g., Kittay 2010, 397; see also Kittay 2017, 25). We can shed a little light on why Kittay hasn't arrived at a settled view of animals and ethics by considering the following, somewhat confusing juncture in her thought. Kittay is clearly committed to the idea that we require moral categories with an essential reference to a conception of what is humanly important in order to bring the worldly life of any human being, however cognitively endowed, into view in ethics (see the text above and also Kittay 2010, 408, and 2017, 28). This commitment ought to have freed her up from the project, decisive for Singer and many others, of providing a theoretical account of the *grounds* of human moral standing, and, as we have seen, she sometimes insists forcefully that she isn't obliged to supply Singer with such grounds. Yet Kittay now says she needs to confront Singer and others with "an alternative" theoretical account of what undergirds the moral standing of human beings, and she appeals in this connection to what she calls "relational properties" of human beings (Kittay 2017, 36). It doesn't seem unreasonable to connect Kittay's new interest in theorizing about these properties with the fact that she doesn't fully explore the philosophically radical aspects of her view that speak against the very idea that we need to theorize about grounds of human moral status. Kittay never specifically asks *why*—as she herself maintains—we require moral categories to get worldly human existence into focus in ethics. Perhaps if she had asked this question, she would have inquired whether we also require moral categories to bring the worldly lives of animals into view in ethics. Perhaps then she would have inquired whether, just as—in her view—humans figure in moral thought as creatures who matter just as the kinds of beings they are, animals of different kinds figure in moral thought as creatures who matter just as the kinds of beings they are. If she had addressed these questions, she would have been more advantageously placed to argue—as she now seems to want to—that we are obliged to jettison the idea of an ethical ranking of life-forms and that, far from being a circumstance we should lament, this better equips us to defend basic human equality.

Abstracting from these involved interpretative issues having to do with Kittay's work, it *is* possible to identify appropriate roles for references to animals in ethical thought about cognitive disability. But this is an indefinitely large topic. One way to approach it would be to investigate the work of people who identify as neuro-atypical and who claim that their particular forms of atypicality give them unusual insight into the lives of animals of specific kinds (see, e.g., Grandin 1995 and Prince-Hughes 2004). Another approach might start from a consideration of the account, laid out above, of how human beings and animals enter moral thought. Encoded in this account is the idea that we require a certain emotional responsiveness or sensibility in order to recognize modes of human and animal expression. This epistemological idea has implications for how in ethics we conceive of the human condition. At issue is a point about how our affective endowments are internal to our rational capacities or, alternately, a point about how certain endowments we have specifically as embodied, fleshly beings are internal to reason. Drawing on this point, we can give a distinctive spin to the now widespread idea that our standing as beings of a rational kind does not exempt us from the contingencies of animal life. We can, that is, give distinctive expression to the idea that—and this is a theme sounded in the work of many great contemporary literary and other artists—as beings of a reasoning kind we are in the condition of animals. To acknowledge this condition is to recognize that there is an ineliminable element of chance both to whether we develop the capacity to reason and to whether we retain it. A sound and illuminating bit of moral thinking might urge us to live in a way that reflects respect for the glorious and terrifying vulnerability that thus binds our lives to the lives of animals—and to at least in this regard accept a role for animal comparisons in ethical thought that touches on questions of cognitive disability.

Acknowledgments

I presented early versions of this chapter at the Disability—Perspectives, Challenges and Aspiration conference sponsored by the Philosophy Department of the University of Tennessee, at the School of Social Science seminar at the

Institute for Advanced Study in Princeton, New Jersey, at the Department of Comparative Thought and Literature at Johns Hopkins University, and at the Philosophy Institute at the Freie Universität in Berlin. I also circulated an early version for discussion at a meeting of the Ira W. DeCamp Bioethics Seminar at Princeton University's University Center for Human Values. I am grateful for the helpful feedback I received on these occasions. I owe particular debts, for their constructive comments, to Susan Brison, Cora Diamond, Jacob Dlamini, Didier Fassin, Lori Gruen, Nathaniel Hupert, Eva Feder Kittay, Nick Langlitz, Paola Marrati, Ally Peabody, Fiona Probyn-Rapsey, Silvia Sebastiani, Peter Singer, and Michael Williams.

Notes

1 There is no suggestion here that animal comparisons *necessarily* degrade human beings. The claim is that certain classic animal comparisons were *in fact* intended to degrade specific groups of human beings. I am grateful to this collection's editors for mentioning the importance of underlining this point.
2 The well-represented position outlined in this paragraph is consistent with holding that an open-ended number of empirical things are in themselves ethically significant. The point is that their ethical significance has to be established by elements of ethical theory or, alternately, by exercises of a strictly practical faculty of reason, and not by exercises of a faculty of reason that, even if partly practical, has an essentially theoretical or world-guided dimension.
3 This is perhaps the right place to acknowledge that some moral thinkers agree with Singer in accepting the constraints of the neutral conception of reason while also insisting on the moral importance of the sheer fact of being human. A stance of this sort is characteristic of Kantian moral philosophers. For relevant discussion of the work of such moral philosophers, see Crary (2018b).

References

Anderson, E. (2014), "Human Dignity as a Concept for the Economy," in M. Duwell et al. (eds.), *Cambridge Handbook of Human Dignity*, 492–497, Cambridge: Cambridge University Press.

Baynton, D. (2001), "Disability and the Justification of Inequality in American History," in P. K. Longmore and L. Umansky (eds.), *The New Disability History: American Perspectives*, 33–57, New York: New York University Press.

Bodey, S. (2017), "My Son Has Autism. Discrimination Almost Cost Him His Life," *Washington Post*, August 30.

Boo, K. (1999), "Forest Haven Is Gone: But the Agony Remains" *Washington Post*, March 14, 1999. Available online: http://www.washingtonpost.com/wp-srv/local/daily/march99/grouphome14.htm (accessed May 22, 2018).

Burleigh, M. (1994), *Death and Deliverance: 'Euthanasia' in Germany c. 1900-1945*, Cambridge: Cambridge University Press.

Crary, A. (2001), "A Question of Silence: Feminist Theory and Women's Voices," *Philosophy* 76: 371–395.

Crary, A. (2002), "What Do Feminists Want in an Epistemology?" in P. O'Connor and M. Scheman (eds.), *Re-Reading the Canon: Feminist Interpretations of Wittgenstein*, 97–118, University Park: Penn State Press.

Crary, A. (2016), *Inside Ethics: On the Demands of Moral Thought*, Cambridge: Harvard University Press.

Crary, A. (2018a), "Ethics," in L. Gruen (ed.), *Critical Terms in Animal Studies*, Chicago: University of Chicago Press.

Crary, A. (2018b), "Cognitive Disability and Moral Status," in A. Cureton and D. Wasserman (eds.), *Oxford Handbook of Philosophy and Disability*, Oxford: Oxford University Press.

Crary, A. (2018c), "The Methodological Is Political," *Radical Philosophy*, 202.

Crary, A. (forthcoming[a]), "Objectivity," in J. Conant and S. Greves (eds.), *Wittgenstein: Basic Concepts*, Cambridge: Cambridge University Press.

Crary, A. (forthcoming[b]), "Seeing Animal Suffering: A Lesson from Tolstoy," in A. Linzey (ed.), *Animal Theologians*.

Darwin, C. ([1871] 1909), *The Descent of Man*, London: John Murray.

Darwin, C. ([1859] 2006), *On the Origin of Species by Means of Natural Selection*, London: Dover.

De Lazari-Radek, K. and Singer, P. (2014), *The Point of View of the Universe: Sidgwick and Contemporary Ethics*, Oxford: Oxford University Press.

Desmond, A. and Moore. J. (2009), *Darwin's Sacred Cause: Race, Slavery and the Quest for Human Origins*, London: Allen Lane.

Diamond, C. (1978), "Eating Meat and Eating People," *Philosophy* 53 (206): 465–479.

Diamond, C. (1991), "The Importance of Being Human," in D. Cockburn (ed.), *Human Beings*, 35–62, Cambridge: Cambridge University Press.

Dombrowski, D. (1997), *Babies and Beasts: The Argument from Marginal Cases*, Chicago: University of Illinois Press.

Drake, S. (2010), "Connecting Disability Rights and Animal Rights—A Really Bad Idea," October 11. Available online: http://notdeadyet.org/2010/10/connecting-disability-rights-and-animal.html (accessed September 15, 2017).

Gaita, R. (2002), *The Philosopher's Dog*, Melbourne: Text Publishing.

Gallagher, H. (1995), *By Trust Betrayed: Patients, Physicians and the License to Kill in the Third Reich*, St. Petersburg: Vandamere Press.

Gelb, S. A. (2008), "Darwin's Use of Intellectual Disability in *The Descent of Man*," *Disability Studies Quarterly* 28 (2).

Gill, M. (2015), *Already Doing It: Intellectual Disability and Sexual Agency*, Minneapolis: University of Minnesota Press.

Grandin, T. (1995), *Thinking in Pictures: My Life with Autism*, New York: Vintage Books.

Hacking, I. (2009a), "Autistic Autobiography," *Philosophical Transactions of the Royal Society* 364: 1467–1473.

Hacking, I. (2009b), "Humans, Aliens and Autism," *Daedalus* 138 (3): 44–59.

Hacking, I. (2010), "How We Have Been Learning to Talk about Autism: A Role for Stories," in L. Carlson and E. F. Kittay (eds.), *Cognitive Disability and Its Challenge to Moral Philosophy*, 261–278, Oxford: Blackwell Publishing Ltd.

Haeckel, E. (1914), *The History of Creation, or the Development of the Earth and Its Inhabitants by the Action of Natural Cause Volume Two*, 6th edition, New York: D. Appleton & Company.

Kateb, G. (2011), *Human Dignity*, Cambridge: Harvard University Press.

Kittay, E. F. (1999), *Love's Labor: Essays on Women, Equality and Dependency*, London: Routledge.

Kittay, E. F. (2005), "At the Margins of Moral Personhood," *Ethics* 116: 100–131.

Kittay, E. F. (2010), "The Personal Is Political Is Philosophical: A Philosopher and Mother of a Cognitively Disabled Person Sends Notes from the Battlefield," in L. Carlson and E. F. Kittay (eds.), *Cognitive Disability and Its Challenge to Moral Philosophy*, 393–413, Oxford: Blackwell.

Kittay, E. F. (2017), "The Moral Significance of Being Human," January 6, Presidential Address, Eastern Division Meeting, American Philosophical Association, Baltimore.

Kulick, D. and Rydström, J. (2015), *Loneliness and Its Opposite: Sex, Disability and the Ethics of Engagement*, Durham: Duke University Press.

Kymlicka, W. (2017), "Human Rights without Human Supremacism," *Canadian Journal of Philosophy*. Available online: DOI: 10.1080/00455091.2017.1386481 (accessed November 17, 2017).

Lovibond, S. (1989), "Feminism and Postmodernism," *New Left Review* 78: 5–28.

Maritain, J. ([1944] 2012), *Christianity and Democracy*, San Francisco: Ignatius Press.

McMahan, J. (2005), "Our Fellow Creatures," *The Journal of Ethics* 9: 353–380.

McMahan, J. (2008), "Challenges to Human Equality," *The Journal of Ethics* 12: 81–104.

McMahan, J. (2009), "Radical Cognitive Limitation," in B. Kimberly and A. Cureton (eds.), *Disability and Disadvantage*, 240–259, Oxford: Oxford University Press.

McMahan, J. (2010), "Cognitive Disability and Cognitive Enhancement," in L. Carlson and E. F. Kittay (eds.), *Cognitive Disability and Its Challenge to Moral Philosophy*, 345–367, Oxford: Blackwell.

Mulhall, S. (2009), *The Wounded Animal: J. M. Coetzee and the Difficulty of Reality in Literature and Philosophy*, Princeton, NJ: Princeton University Press.

Price, M. (2011), *Mad at School: Rhetorics of Mental Disability and Academic Life*, Ann Arbor: University of Michigan Press.

Prince-Hughes, D. (2004), *Songs of the Gorilla Nation: My Journey through Autism*, New York: Harmony Books.

Rachels, J. (1990), *Created from Animals: The Moral Implications of Darwinism*, Oxford: Oxford University Press.

Raffles, H. (2007), "Jews, Lice, and History," *Public Culture* 19 (3): 521–566.

Samuels, E. (2014), *Fantasies of Identification: Disability, Gender, Race*, New York: New York University Press.

Simplican, S. (2015), *The Ability Contract: Intellectual Disability and the Question of Citizenship*, Minneapolis: University of Minnesota Press.

Singer, P. (1990), "Bioethics and Academic Freedom," *Bioethics* 4 (1): 33–44.

Singer, P. (1991), "On Being Silenced in Germany," *The New York Review of Books*, 1–8.

Singer, P. (1992), "A German Attack on Applied Ethics: A Statement by Peter Singer," *Journal of Applied Ethics* 9 (1): 85–91.

Singer, P. (1994), *Rethinking Life and Death: The Collapse of Our Traditional Ethics*, New York: St. Martin's Griffin.

Singer, P. (1999a), "Reply to Michael Huemer," in D. Jamieson (ed.), *Singer and His Critics*, 391–392, Oxford: Blackwell.

Singer, P. (1999b), "A Response," in D. Jamieson (ed.), *Singer and His Critics*, 269–335, Oxford: Blackwell.

Singer, P. (2009), *Animal Liberation*, New York: HarperCollins.

Singer, P. (2010), "Speciesism and Moral Status," in L. Carlson and E. F. Kittay (eds.), *Cognitive Disability and Its Challenge to Moral Philosophy*, 331–344, Oxford: Blackwell.

Spicker, P. (1990), "Mental Handicap and Citizenship," *Journal of Applied Philosophy* 70 (2): 139–151.

Taylor, S. (2017), *Beasts of Burden: Animal and Disability Liberation*, New York: The New Press.

Waldron, J. (2017), *One Another's Equals: The Basis of Human Equality*, Cambridge: Harvard University Press.

Winch, P. (1990), *The Idea of a Social Science and Its Relation to Philosophy*, London: Routledge & Kegan Paul.

Wolfe, C. (2008), "Learning from Temple Grandin, or, Animal Studies, Disability Studies, and Who Comes after the Subject," *New Formations* 64: 110–123.

Wolfensberger, W. (1972), *The Principle of Normalization in Human Services*, Ontario: G Allan Roeher Inst Kinsman.

Chapter 9

THE PERSONAL IS POLITICAL: ORTHOREXIA NERVOSA, THE PATHOGENIZATION OF VEGANISM, AND GRIEF AS A POLITICAL ACT

Vasile Stanescu and James Stanescu

The invention of a new "disease"

"Orthorexia nervosa" is a term that means "fixation on righteous eating" (Bratman 1997). Dr. Steve Bratman, a medical doctor who focuses on holistic treatments, coined the term in the October 1997 issue of the *Yoga Journal*. The *Yoga Journal*—despite having the term "Journal" in the name—is not an academic journal nor does it provide peer-review.[1] However, despite the absence of any peer-reviewed or academic studies on the topic, Bratman subsequently began to diagnose his patients with what, he had determined, was a newly discovered eating disorder. Orthorexia nervosa is not recognized by the American Psychiatric Association, is not included in the *DSM-5* (*Diagnostic and Statistical Manual of Mental Disorders*) or any other medical manual (American Psychiatric Association 2017),[2] and to this day, has few credible peer-reviewed studies to substantiate its existence;[3] it does not represent an actual medical diagnosis. However, it has caught on in the wider popular culture including multiple magazine articles, blog post, books, and television news (Kaplan 2015). For example, largely favorable articles about the condition have appeared in virtually every newspaper in the United States one can think of, including *The New York Times* (Ellin 2009), *The New Yorker*, (Specter 2014), the *Washington Post* (Kaplan 2015), *Newsweek* (Reynolds 2015), and the *Wall Street Journal* (Reddy 2014). Beyond the United States, the phenomenon of orthorexia nervosa also appeared (in just one year) in *The Australian* (2016), on SBS (Verghis 2017), in the *Queensland Times* (Norton 2016), *PerthNow* (Jurewicz 2016), and *Huffington Australia* (Blatchford 2016). Therefore, while not an actual diagnosis,

This chapter was conceived as a partnership with James Stanescu, whose work directly inspired it. Because of starting a new tenure track, James Stanescu was not able to cowrite the chapter, and he felt it was, therefore, wrong to be listed as a coauthor. However, I disagree. Please read everything as coauthored with James Stanescu, as many of the ideas—if not the actual words—came directly from him.

unofficially the "diagnosis" has come to represent a large and growing body of popular knowledge and cultural belief. While the exact criteria are still in flux, overarching indicators of orthorexia supposedly include practices such as: "Violation of self-imposed dietary rules causes exaggerated fear of disease, sense of personal impurity and/or negative physical sensations, accompanied by anxiety and shame" and "Dietary restrictions escalate over time, and may come to include elimination of entire food groups" (Bratman 2017a).

While in theory orthorexia is open to any type of dietary preference, a critical reading of the literature suggests that, in reality, it is primarily focused on pathologizing vegetarianism and veganism as a type of eating "disorder." For example, Bratman was a strict vegan before diagnosing himself with orthorexia (Bratman 1997). Likewise, in the original essay in which Bratman coined the term, while including some other dietary practices, he focused primarily on examples of vegetarian and vegan practices, such as vegetarians who refused to eat vegetables cooked in the same pots as meat,[4] a recovering alcoholic who was made to feel guilty for drinking milk, as well as Bratman's personal examples of relearning to eat Kraft cheese and enjoying a triple scoop of ice cream (Bratman 1997). Bratman recently wrote the foreword to a book entitled *Breaking Vegan,* a memoir from Jordan Younger, a former vegan blogger who diagnosed herself with orthorexia and, as the title suggests, "broke" from her veganism.[5] As far as we can tell, Bratman cites only diets that have at least some type of vegetarian or vegan component, such as a macrobiotic or a raw vegan diet, as dangerous, while explicitly sanctioning as not dangerous other highly restrictive diets that possess a meat-centered component, for example the paleo diet (Bratman 2017a). Such an omission seems odd, since the paleo diets require the complete elimination of food groups such as all cereal grains, all legumes (including peanuts), and even potatoes (Vandyken 2016). In other words, it seems that—at least for Bratman—only diets that eliminate meat, eggs, or dairy run the risk of ever developing into orthorexia.[6]

To be fair, Bratman reassures his readers that it is at least theoretically possible to be a vegan and not suffer from orthorexia. However, it is not clear how that would be possible since, by definition, every vegan would meet several of the Bratman's published criteria for orthorexia, such as elimination of entire food groups, as well as such warning signs as "distress or disgust when in proximity to prohibited foods" or "moral judgment of others based on dietary choices" (Bratman 2017a). In other words, as far as we can tell, the only way anyone could be a vegan and not suffer from orthorexia, under Bratman's interpretation, would be if the person occasionally still consumed meat, diary, or eggs; did not find meat unpleasant to be around; did not criticize or judge the meat-eating practices of others; and did not mind attending social events in which animal products were the only food available. While a person who was occasionally vegan for, say, minor weight loss or health reasons might manage to not to be classified as suffering from orthorexia; it seems likely that every ethical vegan would.

It is important to note the manner in which Bratman mixes his diagnosis of orthorexia, which—as earlier noted—is not actually a disease, with anorexia, a very real and serious medical condition. The similarity begins in the name itself—"orthorexia nervosa" was clearly chosen to sound like anorexia nervosa. Furthermore, the term seems to reclassify the decision to become vegetarian and vegan from an ethical decision—particularly when conducted by young women—to suffering from a dangerous eating disorder paralleled to anorexia. For example, on his website Bratman has a link to a story about Kate Finn who—while medically diagnosed as suffering from anorexia—died, Bratman tells us, because she suffered from orthorexia (Bratman 2003). This account is featured on a website called Beyond Veg, which attempts to convince people not to be vegetarian or vegan, as they see it as inherently unhealthy (Beyond Vegetarianism n.d.). However, they do promote the "paleo diet" that, as earlier mentioned, would seem to meet all of the same criteria of orthorexia (exclusion of entire food groups, etc.). Likewise, it seems clear that Jordan Younger, the ex-vegan blogger discussed earlier, actually suffered from anorexia, a point she herself has admitted, explaining: "I developed obsessions and anxiety around food. It wasn't about veganism. I had restrictions on top of veganism" (James 2015). Indeed, in both of these cases (Jordan Younger and Kate Finn), the health question seems less to do with veganism and far more to do with their desire to eat exclusively raw foods and attempts to engage in—extremely—long juice fasts (Pfeffer 2014), neither of which are standard practices that are part of ethical veganism. Unfortunately, this distinction between ethical veganism on the one hand and not consuming enough calories to maintain health on the other hand is a distinction that not only does Bratman not make clear but, in fact, frequently blurs. For example, on his blog Bratman explains:

A high percentage of the personal communications I receive on this blog involve such young women and their distressed parents. At the fragile physical onset of puberty, these girls are following "healthy food diets" that amount to starvation. A frequent entry point for orthorexia in this group is ethical veganism. This is a philosophy with much to admire and naturally appeals to any idealistic person. Unfortunately, in a young woman this noble impulse can combine with others that are less benign to create an eating disorder. (Bratman 2015)

Anorexia is a real and serious disease, which is caused—in part—by our thin-obsessed and fat-shaming culture. Certain aspects of the animal rights community—for example books such as *Skinny Bitch* (Freedman 2009)—are, therefore, particularly problematic in their use of fat-shaming as a form of vegan promotion. Likewise, it is possible that some people with a preexisting eating disorder—such as anorexia—might claim to be "vegan" in order to hide their eating disorder. And, finally, it is entirely possible that a new or young vegan may not have the necessary information to make wise nutritional decisions. Vegans and animal rights proponents may need to do more to make sure that everyone makes wise nutritional decisions if they transition into a vegan diet. However, none of these are the arguments that Bratman is making. Instead, he is interviewing fears of concerned parents over their daughters' desire to become ethical vegans and linking this decision with preexisting—but unrelated—worries parents may have of their daughters becoming anorexic. Moreover, we are troubled by the way Bratman uses this diagnosis to discount any reasons that a young woman could choose to become an ethical vegan— such as massive animal abuse or environmental degradation—and instead suggests (as he seems to throughout this writing) that the only reason would be a desire for "purity" or an exercise in "control" (Bratman 2015). For example, he explains that young women's "real" desire for going vegan is not ethics—or, indeed, animals at all—but instead the desire to control "the messiness of being human" and to "escape the horrible complexities of life" (Bratman 2015). In any case, Bratman assures us the young female vegan has "practically no idea what she is truly feeling" (Bratman 2015). Indeed, even if a female vegan claims to know why she is choosing veganism, these insights would be irrelevant since she is, in essence, hysterical: according to Bratman, "her brain is malnourished and not working properly" (Bratman 2015).

As Laura Wright writes on this move to pathologize (as anorexia) the ethical reasons behind women's desire to transition to veganism:

> Including animal rights in the discussion about why women choose to become vegetarian or vegan certainly will not alleviate the negative discourse about veganism that pervades both the mainstream media and scientific studies of the links between non-normative diet and eating disorders, but it would certainly allow for more honest analysis, and it might empower instead of pathologize such a choice as having less to do with restricting female diet and more to do with making productive connections between health, feminism, and animal welfare. (Wright 2012)

It is one thing to argue that a concerned parent or society as a whole may wish to make sure that anyone transitioning to veganism has adequate information and support to make informed and healthy nutritional decisions. It is quite another to repeat the preexisting cultural trope that women should simply not be listened to at all because they are hysterical, crazy, diseased, weight-obsessed, and irrational.

Fellowship of the table

Our critique of Bratman's work—as well as the basic idea of orthorexia— is based primarily on the idea of statistical normalization. In other words, under a theory of orthorexia, the main problem with becoming vegan is that it violates established social norms that, in turn, could lead to social isolation. For example, here is the way in which Bratman discusses, in his original article from 1997, how he came to realize that he suffered from orthorexia:

> I pursued wellness through healthy eating for years, but gradually I began to sense that something was going wrong. The poetry of my life was disappearing. My ability to carry on normal conversations was hindered by intrusive thoughts of food. The need to obtain meals free of meat, fat, and artificial chemicals had put nearly all social forms of eating beyond my reach. I was lonely and obsessed. (Bratman 1997)

Vasile Stanescu's research, over the last several years, has focused on the rise of so-called "humane meat" and "locavore" eating movements.[7] What is revealing is the striking similarity between the way in which locavores and proponents of humane meat discount veganism and the way in which Bratman discusses his view of orthorexia. We view this diagnosis of orthorexia as part of a larger discourse that what is wrong with veganism and animal rights is our refusal to "break bread" with others and to socially "outcast" ourselves. In both cases, the argument is that veganism will make one a "social outcast" and the claim that veganism itself is based on a supposed desire for "purity." For example, notice the similarities in the way that Michael Pollan, the most famous advocate for locavorism, describes why he chooses not to give up his own brief experiment with vegetarianism: "[E]ating meat is … also more sociable, at least in a society where vegetarians still represent a relatively tiny minority … What troubles me most about my vegetarianism is the subtle way it alienates me from other people and, odd as this might sound, from a whole dimension of human experience" (Pollan 2006).

Both Bratman and Pollan reject their experiment with vegetarian or veganism because, in their view, it makes them abnormal, statistically strange, that it takes them away from other people, isolates them. Likewise, both of them romanticize eating meat or dairy as "the poetry of my life" in Bratman's case and as "a whole dimension of human experience" for Pollan (Bratman 1997, Pollan 2006). So, too, both argue that what underlies this definable loss of the human experience is a type of infantilism, a type of running away from the messiness of life and fear of our own mortality. In fact, Pollan goes so far as to assert that animal rights represent a new type of "Puritanism" (Pollan 2006).[8]

Nor are these arguments limited to Pollan alone, in fact, they appear—universally—throughout every argument or memoir for "humane" or "local" meat that we have ever read. For example, Catherine Friend, a humane farmer who has published two memoirs, assures her readers that she still eats factory-farmed meat for 25 percent of all of her meals because it would not be compassionate enough for her, or her readers, to become too fanatical even about their "compassionate carnivorism" (Friend 2008, 240).[9] Likewise, Donna Haraway—an important writer in the fields of both feminism and animal studies—claims that vegans fail to understand the messiness and complexities of life, are primarily motivated by a desire for purity, and, therefore, refuse to "break bread" with those who eat meat (Haraway 2007, 294–295).[10] But by far our favorite example is Kathy Rudy who in her text *Loving Animals* asserts that veganism is the same as lesbian separatism—a position she critiques. As she writes: "Veganism is a radical lifestyle change that most of society will never embrace. It's too much like the kind of lesbian separatism that circulated in the 1970's, a radical ideology that felt the world would be a better place without any men in it at all" (Rudy 2013, 104). Perhaps only veganism could manage to be simultaneously compared to both "puritanism" and "lesbian separatism."

Overall, it is odd to hear such appeals to magical surrealism (e.g., "the poetry of life" or "a whole dimension of human experience") in conversations about mental health and ethical decision-making. How exactly is one supposed to argue with logic like that? Do vegans not possess a "fellowship of the table" when we eat together? If fellowship of the table is of such extreme importance, why not simply always serve vegan meals for company, since everyone can always eat vegan? Is a firm commitment to social justice not a desirable part of "the human experience" or the "poetry of life"? What is the most fascinating in all of these magical assertions is that the vegan and animal rights activists are in fact somehow "insane" for being too rational or too logical. Such an assertion represents a fascinating reversal of the stereotype of feminine hysteria. The argument is not that we are crazy because we are too emotional (or—to be exact—not only because we are too emotional), we are pathologized because we have become *too logical*. For example, some of these theorists, such as Rudy, intentionally reference the feminist care ethics tradition of Josephine Donovan and Carol Adams (Donovan 2007), not to argue for animal rights, but, instead, to argue against utilitarian arguments for veganism as being, in her words, "too cold," "clinical," and "sterile" (Rudy 2013, 104). "In other words," she claims, "it lacks heart" (104).[11] Nor is this the only permanent double bind or false choice that vegans find themselves in. If we are not strict enough, we are accused of hypocrisy. If we are, instead, too strict (by whoever's standard) we become insane, fanatical, and obsessed with purity. For example, what we most appreciate about Rudy's comparison to lesbian separatism is the manner in which—on literally the exact opposite page of her book—she indicts vegans for not being pure enough (Rudy 2013, 105). Vegans are, somehow, both too pure and yet never quite pure enough. What exact level of rationality, emotion, or ethical commitment to veganism is acceptable always remains unclear; what is clear is that whatever choices we make or tactics we choose, the vegan—somehow—always gets it wrong.

The vegan as "pervert"

In his College de France lecture *Abnormal*, Michel Foucault charts the manner in which the view of statistical norms become the basis of all forms of madness or insanity (Foucault 2004). Specifically, he lays out how earlier classical ideas of "the natural monster" (which, Foucault suggests, focused primarily on persons who were viewed as transgressive against the "natural laws" of gender and species) become transformed into the "moral monster" who transgresses against the social contract and, finally, transformed into the "abnormal monster" or—as Foucault phrases it—"the pervert." For Foucault, the "pervert" represents someone who fails to match statistical models of normalcy or, one might say, one who violates what might be viewed as "statistical law." As Foucault phrases it: "Psychiatry introduced something that until then was partly foreign to it: the norm understood as a rule of conduct, informal law, and principle of conformity opposed to irregularity, disorder, strangeness, eccentricity, unevenness and deviation … the norm as functional regularity, as the principal of an appropriate and adjusted function: the 'normal' as opposed to the pathological, morbid, disorganized, and dysfunctional" (162). While Foucault's main focus is on issues related specifically to sex—such as fears over homosexuality and childhood masturbation—in providing his genealogy of the way in which all views and beliefs could become pathologized, he mentions of the manner in which even animal rights, when it was first introduced in the nineteenth century, was also pathologized as a type of statistical "disorder." As Foucault writes:

> When, for example, a society for the protection of animals conducts a campaign against vivisection, Magnan, one of the big names in psychiatry at the end of the nineteenth century, discovers a syndrome: the antivivisectionist syndrome. I want to emphasize that, as you can see, there is nothing here that is the symptom of an illness: it is a syndrome, that is to say, a partial and stable configuration referring to a general condition of abnormality. (311)

For Foucault, this reference has the quality of an aside to simply demonstrate the absurd degree that that level of statistical pathologization might reach; however, we wish to take his claims of the pathologization of animal rights quite seriously. For Foucault's overarching argument is to chart that way that the "natural monster," the "moral monster," and the "statistical monster"—far from breaking from one another—build upon one another. For example, Foucault's project, in part, is to chart a genealogy of how the so-called sexual natural monster of gender blending becomes viewed as the moral monster of the sexual criminal, until, ultimately, it becomes solidified into the nineteenth-century discovered disease of "homosexuality." So too, we believe that we can witness a similar continuity in the modern blending of supposed natural, moral, and statistical norms in order to produce the current idea of the morally transgressive vegan who now suffers from the newly discovered "disease" of orthorexia and has become a social pariah (we are told) who disrupts the sacred fellowship of the table. Indeed, the essence of Pollan's argument about how and why veganism is so disruptive invokes, repeatedly, this same type of "natural law" critiqued by Foucault. As Pollan phrases it:

Aristotle speaks of each creature's "characteristic form of life". For domesticated species, the good life, if we can call it that, cannot be achieved apart from humans—apart from our farms and, therefore, our meat eating. This, it seems to me, is where animal rightists betray a profound ignorance about the workings of nature. To think of domestication as a form of enslavement or even exploitation is to misconstrue the whole relationship, to project a human idea of power onto what is, in fact, an instance of mutualism between species. Domestication is an evolutionary, rather than a political, development. (Pollan 2006, 320)

There is far too much to unpack in this paragraph here; however, the essential point we want to take away is the idea that animal rights betray a profound ignorance of the "workings of nature" (i.e., natural law) and, likewise, a profound ignorance of a type of "species contract," (i.e., social contract) leading to a criticism based on statistical irregularity (i.e., veganism becomes "unnatural" only because it is uncommon). We see in Pollan and Bratman, via Foucault, the same leitmotif of the vegan, and in particular, the animal rights activist as a natural, moral, and statistical "monster." Much like Foucault's insight of causes of the pathogenization of homosexuality, what disturbs Pollan so much in the move toward veganism is its perversion of what he views as "natural law" and arrangement of species hierarchy and the "social contract" between humans and domesticated animals. In other words, for Pollan, the vegan is a "pervert." They have perverted the natural rule of man over animal, the moral rule of the social contract between human and other species, and perverted the statistical—yet still sacred—fellowship of table and, therefore, perverted the very essence of the "human experience." If one hears in such claims a certain similarity to arguments for the "sacredness of marriage" or a panic against transfolk using a restroom, we should view such similar refrains as far from purely coincidental. In all such cases, there is a conflation of the natural and moral with the merely statistical, tied to a fear of perversion coupled with a desire for "social quarantine." Utilizing Foucault's insights, we must see these statistical claims of social isolation in relation to fears of contagion of social perversion and fears of a transgressive "monster." By which we mean, the seeming objective claims that vegans are a statistical rarity is, in fact, the way both Pollan and Bratman seem to believe that the vegan should be treated and indeed "cured." Not only that we are socially stigmatized and ostracized but in fact that we should be. And thus via Foucault, we can begin to answer back our earlier hypothetical question of why it is that both Pollan and Bratman—in their discussion of madness—utilize such metaphysical language as "the poetry of life" and "a whole dimension of the human experience." And why they seem incapable of imagining a pleasure of eating together, either between vegans themselves or with omnivores, in which everyone is consuming food free of animal products. Such a sociality is unthinkable for either Pollan or Bratman because it fails the necessary quarantine of veganism; such sociality might even represent a further expansion of the perversion of veganism they wished to domesticate in the first place. And hence also this peculiar emphasis that the "cure" of veganism or orthorexia is a speech act; as Foucault makes clear throughout *Abnormal*, the cure for insanity was always the model of the confessional, that speech act that disavowed all the different forms of insanity. Since what was at stake was the pervasion of the social order, what mattered were not the private beliefs as much as the public speech act. Hence, too, we can begin to answer our earlier question of how Bratman believes that a vegan could still eat a vegan diet and yet, somehow, not suffer from orthorexia, even as all criteria would suggest that they must. The solution would seem to be that it is acceptable—in Bratman's view—for someone to refrain from eating animal products as long as they refuse to engage in the speech acts identifying as veganism. What is pathologized is less the reality of the diet—whatever it might be—than the speech act itself of identifying as vegan that "isolates" one. In other words, one can refrain from eating animals only as long as no one knows, as long as one repeatedly disavowals any identity or judgment. Judith Butler has written on the unique nature of the "Don't Ask, Don't Tell" compromise in the US military, where it was not the action, not even the desire, but the speech act of identifying as queer that uniquely criminalized and regulated (Butler 1997, 82). So, too, we want to see an attempt in both locavores and orthorexia to pathologize not the eating of meat—one way or another—but the speech act of identifying as ethically vegan.

To provide a concrete example, while in graduate school, Vasile met a prominent environmental researcher who, through his research, had determined that the consumption of animal flesh was a leading cause of climate change and preventable harm against the environment. And so, he was strictly vegan, except for a single meal of meat he had one day a year. We have long wondered about the utility of that single meal of flesh: if he felt that eating meat possessed some unique nutrient, a single meal a year would be irrelevant. Even in terms of pleasure the connection seems unclear: What possible pleasure could one derive from a single meal a year, particularly when his body had, in all likelihood, lost the ability to even fully digest animal flesh?[12] We finally determined that the decision entirely came down to one of identity. When Vasile first congratulated the researcher on being vegan, he stated that he was not and mentioned his one meal a year as his proof. For him, it seemed, that single meal was how he still proved his "virile" and carnivorous identity; his one meal a year proved that he was, by definition, still not a vegan. In other words, he was not a pervert; he was still socially acceptable at the fellowship of the table. So, too, we have discovered the same tendency in all forms of locavorism and so-called "compassionate" or "humane" meat. Since less than 1 percent of animal products are even marketed as humane meat,[13] many advocates for humane meat, such as Catherine Friend, simply eat factory-farmed meat as a large part of their diet. However some, such as Pollan, regularly insist that since humane meat is so hard to locate, they, in reality, eat almost no meat at all. If what they are telling is true, we must ask what is behind these extremely small and rare bites of so-called "humane" meat that they do eat? As Pollan himself writes: "[W]hy am I working so hard to justify a dinner menu?" (Pollan 2006). If they are, as they claim, "basically" vegetarian already, what is at stake for them in not becoming *actually* vegetarian? We believe the issue is entirely one of identity, no matter how little or how rare their consumption of animals has become. As long as they still eat some meat, any meat, they're not perverted, they're not crazy, and, most importantly for someone such as Pollan, they have not disrupted the natural or moral order, and, arguably, they should still be welcomed at the table of fellowship of (straight?) men. As Chloë Taylor has argued on this same connection:

> I will argue that alimentary appetites, like sexual appetites, continue to be sites of normalization, or that how we eat is a target of what Foucault calls disciplinary power. Moreover ... just as the sexual and alimentary monsters were frequently fused in the late eighteenth- and early nineteenth-century popular imaginary, so today the abnormalities of eating and sex are often conflated, with male vegetarians in particular suspected of being—queer. The normalization of sex and eating are thus not only analogous but inter-related and mutually reinforced. (Taylor 2012, 132)

Before you decide that we are making too much of this connection between Pollan's hostility to veganism and sex, it is the exact analogy that Pollan himself makes in the same section where he discusses veganism as "a loss of the humane experience." As he phrases it: "We should at least acknowledge that the human desire to eat meat is not, as the animal rightists would have it, a trivial matter, a mere gastronomic preference. By the same token we might call sex—also now technically unnecessary for reproduction—a mere recreational preference. Rather, our meat eating is something very deep indeed" (Pollan 2006).

Grief as political act

There is in fact a feeling of insanity in our experience of being vegan, but it is not the insanity of orthorexia; it is instead the insanity that comes from the feeling of being surrounded by a "banality of evil." As J.M. Coetzee writes via the character of Elizabeth Costello:

> I seem to move around perfectly easily among people, to have perfectly normal relations with them. Is it possible, I ask myself, that all of them are participating in a crime of stupefying proportions? Am I fantasizing it all? I must be mad! Yet every day I see the evidence. The very people I suspect produce the evidence, exhibit it, offer it to me. Corpses. Fragments of corpses that they have bought for money. (Coetzee 1999, 69)

Or, as James Stanescu has written in a similar vein:

> To tear up, or to have trouble functioning, to feel that moment of utter suffocation of being in a hall of death [the meat aisle in the grocery store] is something rendered completely socially unintelligible. Most people's response is that we need therapy, or that we can't be sincere. So most of us work hard not to mourn. We refuse mourning in order to function, to get by. But that means most of us, even those of us who are absolutely committed to fight for animals, regularly have to engage in disavowal. (Stanescu, J. 2012a, 568)

As Dinesh Wadiwel has recently argued, the correct analogy we should use for our understanding of the mistreatment of animals is not individual cruelty, but instead that of war (Wadiwel 2015). For the simple truth is that to torture and kill 70 billion animals each and every year is not the act of one or two individuals of human cruelty, no matter how depraved; it is the work of millions and millions of humans all working in concert. It is a collective action requiring a collective solution. In such a system, to be an ethical vegan is to violate the deepest held, if unspoken, dictate of our entire society. It is to become a conscientious objector—indeed a traitor—to a global war effort toward human supremacy. This is why, we have always felt, we see the constant invocation of the "analogy" to the holocaust throughout animal rights and critical animal studies literature by Coetzee (1999)—even Jacques Derrida (2008)—and so many others (see Adorno 2005, Dujack 2003, Kupfer-Koberwitz n.d., Netz 2004, Patterson 2002, and Singer 1980). Not that any of them are making moral comparisons between incomparable instances, but because it is the only framework they have to invoke that feeling they possess of fighting against a massively organized system of a "banality of evil" (Arendt 2006, 276) that only they seem able to even see.[14]

If Wadiwel is correct and violence against animals is a collective action, requiring a collective solution, what we must learn to do is to transform these individual moments of grief into political acts of resistance. To refuse food when offered by our family members or to grieve the dead bodies that surround us in the meat aisle of a grocery store or a restaurant is, in our current society, to be "crazy," too feminine, and "hysterical," regardless of the gender of the person grieving. But what if it were not so? We believe that Foucault, via Wadiwel's rereading of him, offers us a way to affectively respond to this pathologization of veganism. If what is occurring is war, we must learn to demonstrate that our individual grief represents a collective political action. We see as the best example of this transformation of what was previously viewed as a purely personal grief into a political action is the work of ACT UP. ACT UP represents a direct action advocacy group that through public actions, such as "die-ins," sought to transform the personal grief of AIDS victims and their loved ones (primarily found within the LGBTQAAI+ community) into public and political project.[15] We see in the actions of vegan activists' recent move to host a funeral within the isles of grocery stores as a movement toward highlighting the banality of evil that surround us at all times, as a way to reject actions of personal disavowal, and as a way to transform our own individual moment of grief into effective calls for political change and action. As Lori Gruen has convincingly argued in this same context: "Creating communal possibilities for mourning our companions whom we have loved and lost as well as all of the animals for whom we grieve can take grief out of the closet, out of the realm of the comic or crazy, and make the lives and deaths of animals visible and meaningful" (Gruen 2014).

Carol Hanisch wrote in her essay "The Personal Is Political" (1970) on the tendency to view women's consciousness raising as a type of therapy:

> The very word "therapy" is obviously a misnomer if carried to its logical conclusion. Therapy assumes that someone is sick and that there is a cure, e.g., a personal solution. I am greatly offended that I or any other woman is thought to need therapy in the first place. Women are messed over, not messed up! We need to change the objective conditions, not adjust to them. Therapy is adjusting to your bad personal alternative. (Hanisch, 76)

A view that leads to her conclusion that "One of the first things we discover in these groups is that personal problems are political problems. There are no personal solutions at this time. There is only collective action for a collective solution" (76). The same is true, we would suggest, for our feeling of grief as a vegan and the current attempts to pathologize our political resistance as a new type of personal hysteria requiring therapy. We, too, need to change the objective conditions, not adjust to them. We, too, must come to see that there are no personal solutions to the exploitation of animal suffering—not even personally going vegan alone; there is and can only be our collective action for animal rights in order to achieve a collective solution to our species' war against all other nonhuman animals. Like the feminist, queer, and other social justice movements before us, what animal rights activist need is not therapy, but *change*.

Acknowledgments

We would like to acknowledge Chloë Taylor who first suggested the idea of writing about *orthorexia nervosa* to us and whose work directly helped inspire this chapter.

Notes

1   For example, the *Journal* includes the claim—directly alongside Dr. Bratman's article, "Reality Check: What Is Everything a Part of and Yet Not Part of Anything?

The Source of All Energies! Science Calls It Tachyon." Advanced Tachyon Technologies, "Reality Check," *Yoga Journal* 1997. September/October: 49. While not entirely clear, I think this might be an advertisement (as it includes an "800" telephone number and an offer of a "free catalogue"). It is possible that the *Yoga Journal* holds its articles to a different standard than its advertisements. However, it is clearly not an academic or peer-reviewed journal.

2 See also Kaplan (2015).

3 Since 1997, a small number of peer-reviewed studies on the disease have been published; however, a meta-analysis found "methodological problems in these studies, such as the use of non-validated assessment instruments, small sample size and sample characteristics, which make generalization of the results impossible" (Varga et al. 2013, 103). Likewise, in her dissertation specifically on the question of whether orthorexia should be considered as a valid diagnosis, Erin Michelle Mcinerney-Ernst found "The vast majority of ON [Orthorexia nervosa] research has been focused on very limited samples, such as homogeneous groups of medical students, nutritional students, or performance artists, and lacks experimental rigor" (Mcinerney-Ernst 2011, 26).

4 "The main entree was always vegetarian. However, a small but vocal group insisted on an optional serving of meat. Since many vegetarians would not eat from pots and pans contaminated by fleshly vibrations, this meat had to be cooked in a separate kitchen. The cooks also had to satisfy the Lacto-ovo-vegetarians, or Vegans, who eschewed all milk and egg products" (Bratman 1997).

5 The memoir is a particularly clear example of the way the discourse around orthorexia circulates, as the title was not "Breaking Orthorexia" but instead "Breaking Vegan" (Younger 2016). See also Bratman (n.d.).

6 Note: Other media outlets have used orthorexia to discuss non-vegetarian-type diets (e.g., avoiding gluten) and some have even criticized the paleo diet; these comments are specific to Bratman.

7 See Stanescu 2017, 2016, 2013, 2009, and Jenkins 2014.

8 "A deep current of Puritanism runs through the writing of the animal philosophers, an abiding discomfort not just with our animality, but with the animals' animality too. They would like nothing better than to airlift us out from nature's 'intrinsic evil'—and then take the animals with us. You begin to wonder if their quarrel isn't really with nature itself" (Pollan 2006, 322). See also Stanescu (2012b).

9 See also Stanescu (2014).

10 In *When Species Meet*, Haraway writes a replicate of the arguments made by both Bratman and Pollan about the social isolation of veganism and her view on the possible dangers of extremism. For example, in justifying her decision to eat a feral pig, Haraway explains: "There is a third and last parting bite necessary to explore how to proceed when species meet. No community works without food, without eating *together*... This is a deeply unsettling fact if one wants a pure diet ... There is no way to eat and not to kill, no way to eat and not to become with other mortal being to whom we are accountable, no way to pretend innocence and transcendence or a final peace ... Further, one must actively cast oneself with some ways of life and not those who eat differently to a subclass of vermin, the underprivileged. Or the unenlightened; and giving up on knowing more,

including scientifically, and feeling more, including scientifically, about how to eat well—together" (Haraway 2007, 295).

11 "There are aspects of care theory that deeply resonate with the kind of 'connection' I call for in this book. While to some extent I quarrel with both the foundations of care theory (briefly that women are essentially more caring creatures) and with the outcome of care theory (that animals are not ours to use), the impulse behind feminist care ethics is very attractive because it places the weight of social change in the affective, emotive register (rather than in rationality). I hope to revisit these questions in my next project … Utilitarian can be cold, clinical, and sterile. In other words, it lacks heart" (Rudy 2013, 104).

12 To what degree vegetarians lose the ability to digest meat is a widely debated issue. The overall consensus is that vegetarians do lose the ability to fully digest meat in the short run; however, this ability returns in the long term if vegetarians continue to eat meat. As such, someone who only ate meat once a year would—in all likelihood—experience physical discomfort. For a fuller discussion on the topics, see Lane (2007) and Santillano (n.d.).

13 As Farm Forward, an organization that actually works to support "humane" farms, admits: "[T]he reality of meat is unambiguous. And at Farm Forward we don't pull any punches when we face inconvenient realities: Most of the animals raised and killed for food (more than 99 percent, to be precise) come from unsustainable and cruel factory farms or, in the case of sea animals, other industrial operations.… Every person who adopts a vegetarian diet reduces suffering and environmental degradation *and* helps stretch the small supply of non-factory meat, dairy, and eggs currently available for those who choose to eat meat. As long as the demand for non-factory animal products exceeds the supply to this degree, it is best to avoid even these products. But whatever our approach to eating ethically, the important point to remember is that withdrawing our financial support from factory farming reduces the greatest barrier to a humane, sustainable agriculture: the wealth and power that the factory farm industry draws from the money we funnel to it daily" (Food Choices 2011).

14 We mean banality of evil in the way it was originally framed by Hannah Arendt, as a system whereby it is hard to know your actions are even evil anymore. As she explained, "The trouble with Eichmann was precisely that so many were like him, and that the many were neither perverted nor sadistic, that they were and still are, terribly and terrifyingly normal. From the viewpoint of our legal institutions and of our moral standards of judgment this normality was much more terrifying than all the atrocities put together for it implied… that this new type of criminal, who is in actual act *hostis generis humani*, commits his crime—under circumstances that make it well-nigh impossible for him to know or to feel that he is doing wrong" (Arendt 2006, 276).

15 ACT UP quotes on its website this explanation of how personal grief motivated Charles Clifton, executive director of Test Positive Aware Network in Chicago, into his work for ACT UP: "I'm tired of grieving. I'm tired of remembering. I'm tired of wondering. I'm tired that I still grieve the death of Antonio, who died 15 years ago on October 8th. I'm tired of marking the anniversary of his death. I'm tired of wondering of what might have been. I'm tired of hoping. I'm tired of coping. I'm tired of dates that always remind me of how tired I am. I'm tired of wondering what's next, who's next. I'm tired of this road" (Clifton 2001).

References

Adorno, T. (2005), *Minima Moralia: Reflections on a Damaged Life*, New York: Verso.

American Psychiatric Association (2017), *Diagnostic and Statistical Manual of Mental Disorders*, 5th edition, Arlington, VA: American Psychiatric Publishing.

Arendt, H. (2006), *Eichmann in Jerusalem: A Report on the Banality of Evil*, New York: Penguin.

Beyond Vegetarianism (n.d.), "Beyond Vegetarianism." Available online: http://www.beyondveg.com/ (accessed February 21, 2018).

Blatchford, E. (2016), "Orthorexia Nervosa: The Darker Side of Clean Eating," *HuffPost Australia*, January 3. Available online: http://www.huffingtonpost.com.au/2016/03/01/orthorexia-nervosa_n_9344844.html (accessed February 21, 2018).

Bratman, S. (1997), "Health Food Junkie," *Yoga Journal* (September/October), pp. 42–50.

Bratman, S. (2003), "Fatal Orthorexia." Available online: http://www.orthorexia.com/original-orthorexia-essay/fatal-orthorexia/ (accessed February 21, 2018).

Bratman, S. (2015), "Adolescent Girls and Orthorexia," July 19. Available online: http://www.orthorexia.com/adolescent-girls-and-orthorexia/ (accessed February 21, 2018).

Bratman, S. (2017a), "Healthy Eating vs. Orthorexia," March 26. Available online: http://www.orthorexia.com/healthy-eating-vs-orthorexia/ (accessed February 21, 2018).

Bratman, S. (2017b), "The Authorized Bratman Orthorexia Self-Test," June 8. Available online: http://www.orthorexia.com/ (accessed February 21, 2018).

Bratman, S. (n.d.), "Foreword to Breaking Vegan." Available online: http://www.orthorexia.com/foreward-to-breaking-vegan/ (accessed February 21, 2018).

Butler, J. (1997), *The Psychic Life of Power: Theories in Subjection*, Stanford, CA: Stanford University Press.

Clifton, C. (2001), "I'm Tired of Grieving; I'm Tired of Remembering; Tired of Wondering," *Positively Aware*, November/December. Available online: http://www.actupny.org/ (accessed February 21, 2018).

Coetzee, J. M. (1999), *The Lives of Animals*, A. Gutmann (ed.), Princeton, NJ: Princeton University Press.

Derrida, J. (2008), *The Animal That Therefore I Am*, M. L. Mallet (ed.), Wills, D. (trans.), New York: Fordham University Press.

Donovan, J. A. (2007), *The Feminist Care Tradition in Animal Ethics: A Reader*, New York: Columbia University Press.

Dujack, S. R. (2003), "Animals Suffer a Perpetual 'Holocaust,'" *The Los Angeles Times*, April 21. Available online: http://articles.latimes.com/2003/apr/21/opinion/oe-dujack21 (accessed February 21, 2018).

Ellin, A. (2009), "What's Eating Our Kids? Fears about 'Bad' Foods," *New York Times*, February 25. Available online: http://www.nytimes.com/2009/02/26/health/nutrition/26food.html (accessed February 21, 2018).

Farm Forward (2011), "Food Choices," *Farm Forward*, May 17. Available online: http://www.farmforward.com/farming-forward/food-choices (accessed February 21, 2018).

Foucault, M. V. (2004), *Abnormal: Lectures at the Collège De France, 1974–1975*, New York: Picador.

Freedman, R. K. (2009), *Skinny Bitch; Skinny Bitch in the Kitch*, Old Saybrook: Tantor Media.

Friend, C. (2008), *The Compassionate Carnivore, or, How to Keep Animals Happy, Save Old MacDonald's Farm, Reduce Your Hoofprint, and Still Eat Meat*, Philadelphia, PA: Da Capo Lifelong.

Gruen, L. (2014), "Facing Death and Practicing Grief," in C. J. Adams (ed.), *Ecofeminism: Feminist Intersections with Other Animals and the Earth*, New York: Bloomsbury Publishing.

Hanisch, C. (1970), "The Personal Is Political," in S. F. Koedt (ed.), *Notes from the Second Year: Women's Liberation*, 76–83, New York: Radical Feminism.

Haraway, D. J. (2007), *When Species Meet*, Minneapolis: University of Minnesota Press.

"Hard to Swallow: Is Clean Eating Really Good for You, or a Damaging Obsession?" (2016), *The Australian*, June 10. Available online: https://www.theaustralian.com. au/life/weekend-australian-magazine/clean-eating-and-orthorexia--the-perils-of-maca-root-cacao-nibs-spirulina-chia-seeds-and-other-healthy-food/news-story/02 a1094bbe4e6eb22b60a4d1e0e47a4a (accessed February 21, 2018).

James, S. D. (2015), "Blogger Jordan Younger Reveals How Extreme 'Clean Eating' Almost Killed Her," *Today*, November 11. Available online: https://www.today. com/health/breaking-vegan-author-jordan-younger-confesses-dirty-secrets-clean-eating-t55086 (accessed February 21, 2018).

Jenkins, V. S. (2014), "'One Struggle' in Defining Critical Animal Studies," in A. J. Nocella II, J. Sorenson, K. Socha, and A. Matsuoka (eds.), *Counterpoints: Studies in the Postmodern Theory of Education*, 74–89, New York: Peter Lang Publishing.

Jurewicz, J. (2016), "Orthorexia: Why the Too Clean Diet Spread by Social Media Is Pure Hell," *PerthNow*, May 1. Available online: https://www.perthnow.com.au/ lifestyle/stm/orthorexia-why-the-too-clean-diet-spread-by-social-media-is-pure-hell-ng-0c7e5bc3ca2ea7970a882e09220646d0 (accessed February 21, 2018).

Kaplan, S. (2015), "Psychiatry Doesn't Recognize "Orthorexia"—An Obsession with Healthy Eating. But the Internet Does," *Washington Post*, November 5. Available online: https://www.washingtonpost.com/news/morning-mix/wp/2015/11/05/ psychiatry-doesnt-recognize-orthorexia-an-obsession-with-healthy-eating-but-the-internet-does/?utm_term=.3d32e90d09f4 (accessed February 21, 2018).

Kupfer-Koberwitz, E. (n.d.), *Dachau Diaries*, Chicago: University of Chicago Library.

Lane, A. (2007), "Do Vegetarians Lose the Ability to Digest Meat?" *Chowhound*, November 28. Available online: https://www.chowhound.com/food-news/54220/ do-vegetarians-lose-the-ability-to-digest-meat/ (accessed February 21, 2018).

Mcinerney-Ernst, E. M. (2011), "Orthorexia Nervosa: Real Construct or Newest Social Trend: A Dissertation in Psychology," Ph.D. diss., University of Missouri-Kansas City, Kansas City.

Netz, R. (2004), *Barbed Wire: An Ecology of Modernity*, Middletown: Wesleyan University Press.

Patterson, C. (2002), *Eternal Treblinka: Our Treatment of Animals and the Holocaust*, New York: Lantern Books.

Pfeffer, S. E. (2014), "Popular Food Blogger–the Blonde Vegan–Admits to Eating Disorder," *People*, July 15. Available online: http://www.people.com/article/blonde-vegan-jordan-younger-blogger-eating-disorder-orthorexia (accessed February 21, 2018).

Pollan, M. (2006), *The Omnivore's Dilemma: A Natural History of Four Meals*, New York: Penguin Press.

Reddy, S. (2014), "When Healthy Eating Calls for Treatment," *The Wall Street Journal*, November 10. Available online: https://www.wsj.com/articles/when-healthy-eating-calls-for-treatment-1415654737 (accessed February 21, 2018).

Reynolds, R. C. (2015), "When Eating Healthily Becomes a Fixation," *Newsweek*, April 4. Available online: http://www.newsweek.com/when-eating-healthily-becomes-fixation-319453 (accessed February 21, 2018).

Rudy, K. (2013), *Loving Animals: Toward a New Animal Advocacy*, Minnesota: University of Minnesota Press.

Santillano, V. (n.d.), "Can Eating Meat Really Make Vegetarians Sick?" *More Lifestyle*. Available online: http://www.more.com/lifestyle/exercise-health/can-eating-meat-really-make-vegetarians-sick (accessed February 21, 2018).

Singer, I. B. (1980), *The Seance and Other Stories*, New York: Farrar, Straus and Giroux.

Specter, M. (2014), "Against the Grain: Should You Go Gluten-Free?" *New Yorker*, November 3. Available online: https://www.newyorker.com/magazine/2014/11/03/grain (accessed February 21, 2018).

Stanescu, J. (2012a), "Species Trouble: Judith Butler, Mourning, and the Precarious Lives of Animals," *Hypatia* 27 (3): 567–582.

Stanescu, J. (2012b), "Towards a Dark Animal Studies: On Vegetarian Vampires, Beautiful Souls, and Becoming-Vegan," *The Journal of Critical Animal Studies* 10 (3): 26–50.

Stanescu, V. (2009), "Green Eggs and Ham: Michael Pollan, Locavores, and the Myth of Environmentally Sustainable Meat," *The Journal of Critical Animal Studies* 3, 18–55.

Stanescu, V. (2013), "Why 'Loving' Animals Is Not Enough: A Response to Kathy Rudy, Locavorism, and the Marketing of 'Humane' Meat," *The Journal of American Culture* 36, 100–110.

Stanescu, V. (2014), "Crocodile Tears: Compassionate Carnivores and the Marketing of 'Happy Meat'," in John Sorenson (ed.), *Critical Animal Studies: Thinking the Unthinkable*, 216–233, Toronto: Canadian Scholars Press.

Stanescu, V. (2016), "Beyond Happy Meat," in Donaldson, B. and Carter, C. (eds.), *The Future of Meat without Animals*, 133–154. Lanham, MD: Rowman and Littlefield.

Stanescu, V. (2017), "New Weapons: 'Humane Farming,' Biopolitics, and the Commodity Fetish," in David Nibert (ed.), *Animal Oppression and Capitalism*, 209–228, Santa Barbara: Praeger.

Taylor, C. (2012), "Abnormal Appetites: Foucault, Atwood, and the Normalization of an Animal Based Diet," *Journal for Critical Animal Studies* 10 (4), 130–148.

Vandyken, P. (2016), "What to Eat on the Paleo Diet," *Paleo*, October 12. Available online: http://www.paleo.com (accessed February 21, 2018).

Varga, M., Dukay-Szabó, S., Túry, F., and Furth Eric, F. (2013), "Evidence and Gaps in the Literature on Orthorexia Nervosa," *Eating and Weight Disorders–Studies on Anorexia, Bulimia and Obesity* 18 (2): 103–111.

Verghis, S. (2017), "Orthorexia: When Your Desire for 'Clean Eating' Turns Obsessive," *Special Broadcasting Service*, April 13. Available online: https://www.sbs.com.au/topics/life/health/article/2017/04/13/orthorexia-when-your-desire-clean-eating-turns-obsessive (accessed February 21, 2018).

Wadiwel, D. (2015), *The War against Animals*, Boston, Brill/Rodopi.

Wright, L. (2012), "The Vegetarian/Vegan Anorexia Connection," *The Vegan Body Project*, September 12. Available online: http://veganbodyproject.blogspot. com/2012/09/the-vegetarianvegan-anorexia-connection.html (accessed February 21, 2018).

Younger, J. (2016), *Breaking Vegan: One Woman's Journey from Veganism, Extreme Dieting, and Orthorexia to a More Balanced Life*, Beverly, MA: Fair Winds.

Chapter 10

WOMEN, ANXIETY, AND COMPANION ANIMALS: TOWARD A FEMINIST ANIMAL STUDIES OF INTERSPECIES CARE AND SOLIDARITY

Heather Fraser and Nik Taylor

This is Pip, our 4 year old English Staffy. I cannot put into words how much we love her. She is constantly making us laugh—online respondent.

—Fraser and Taylor, 2015–16

This chapter focuses on two projects where participants were invited to represent their connections to animals through photos, art, poetry, or prose. The first was an online visual project, *Whatisitaboutanimals.com,* and the second project, called *In Good Company,* involved three focus groups where participants talked about their connections to companion animals and shared photographs of them. As elaborated later, both projects were open to all adults; however, it was women who chose to participate. According to the women, such relationships provided love, support, and friendship, and for many, allowed them to experience themselves in more positive and expansive ways. An important subtheme included companion animals soothing women's anxiety. On face value, these findings fit with the burgeoning growth of stories about the benefits diverse groups of humans enjoy from "keeping pets" (also see Taylor et al. 2017). However, from a more critical perspective, what might the women's stories about companion animals mean?

As this chapter explains, a different light is cast over the women's reports of human–animal relationships if read through a feminist animal studies perspective. Pointed questions are raised about power, privilege, and oppression, not just for women differentially located across class, sex, sexuality, race/ethnicity, ability, and age but also for companion animals. A critique of speciesism shines a different light over discussions of "pet-keeping"—a term that assumes humans can rightfully keep animals (captive) as their playthings. It acknowledges that elements of control are evident in human relations with animal companions. Despite this, the chapter ends with a discussion of whether servitude is an inevitable feature of all human–companion animal relationships, considering the possibility of interspecies alliances, specifically alliances of care and solidarity.

The projects: (1) In Good Company *and (2)* Whatisitaboutanimals.com?

Three focus groups were conducted for the *In Good Company: Women and Animal Companions* project, with a total of twenty-five women participants; two groups were held on a university campus and one in a local community organization. For the online "What Is It about Animals?" an open and very broad invitation was made to people over the age of sixteen to post images, video, poems, stories, and other texts to a dedicated website about what their animals meant to them. Curious about what a call would bring, we made no mention of anxiety, depression, or well-being, instead simply telling participants that "we want to know how you experience animals you consider important; how you describe and feel about these relationships." We also made no mention of animal rights other than implicitly, through the instructions for participants to only submit "respectful posts," giving them the power to interpret what that might mean. With 200,000 views from 28,000 unique visitors, we ended up with a total of ninety-four image posts, sixty-eight of which also included text—almost all from women. Some excerpts are outlined in the discussion below.

In both projects the call for participants was open, across gender, class, role, and rank (for the focus groups conducted at the university and for the group held in a community organization). That women overwhelmingly stepped forward for both projects is a wider reflection of women's greater ease, interest in, and openness to talking about animals—particularly companion animals—in public settings, a discussion returned to later in the chapter.

During the *In Good Company* focus groups, we witnessed women not just talking to each other about animals, but bonding with each other (and us) over companion-animal talk. Within each hour of the group discussions, so many emotions were expressed, and among many who had never previously met. Laughter, joy, and shared discussions of funny, poignant, sad, sometimes gut-wrenching stories were narrated. Little was needed from (us) the facilitators, to keep the discussions moving. In each of the three focus groups, women who did not know each other before the groups seemed to talk as easily as those who did, flashing photos of their (animal) loved ones via their phones and in one case, a large iPad, brought in without prompting. For all their differences in age, ethnicity, and class (including role and rank), the women in all three focus groups shared the view that "pets" were not objects or property to be dominated, but family members and cherished friends requiring respectful and sensitive treatment. The same can be said for the tenor of the messages posted online for the *Whatisitaboutanimals.com* project.

Fun, love, and acceptance

Whether online or through the focus groups, participants kept reiterating how much fun, love, and friendship they felt toward their companion animals. Without being directed, they talked about how their companion animals so attentively listened to and comforted them; how they accepted them without judgment or criticism; and how they provided an anchor in hard times.

> This is Millie [dog]. I hate being away from her even if it is for a short amount of time, and I know that feeling is mutual ... She puts my mind at ease and calms me in just about any situation—focus group respondent. (Fraser and Taylor 2015–16)

> For someone that has had a trying life, my Winky [cat] was my one and only reliable constant. Our relationship constantly surprises people as it is clear that we both care about each other conditionally. I would be lost without her in my life—she is my rock!—focus group respondent. (Fraser and Taylor 2017)

> Bessie [dog] very recently passed away ... She had her own Facebook page and was loved by people all around the world ... so lost without her in my life—focus group respondent. (Fraser and Taylor 2017)

> I love her with all my heart, and without her my life wouldn't be the same. I can have the worst day at work, coming home to my Daisy [dog], who is always full of love, hugs and big licks makes everything okay, and instantly better—focus group respondent. (Fraser and Taylor 2017)

Heartfelt accolades continued for the individually named and described companion animals the women participants spoke about, in both our focus groups and online projects. Over and again participants stressed the power of the physical affection shown by companion animals, as well as their acceptance without regard to human preoccupations, such as beauty, status, and money.

> I love that animals are so unselfconscious, so uncontrived, yet sentient and affectionate. They just "are"!—focus group or online respondent. (Fraser and Taylor 2015–16)

> Louis [dog] has been my best friend from the moment I got him. I moved to a new place far from any family or friends. And that's when I adopted Louis ... Even though he's cheeky sometimes, he still gets me through the hard days, and for that I am so thankful.—online respondent. (Fraser and Taylor 2015–16)

Time and again respondents told of how their companion animals got them through "hard times," offering unconditional regard and love, positive reinforcement, and a shoulder to lean on, "I feel blessed having my furry companion (dog) by my side every day, where-ever I go he goes usually. I suffer with a little anxiety and stress out but I'm always less stressed when he's around. No words needed, yet I have the best relationship with my dog" (Fraser and Taylor 2015–16).

Repeated references were made to how companion animals soothed their anxiety and depression, with a few women disclosing their experiences of psychiatric ill-health:

> I look after pets for a living and my life couldn't be better because of it. I have struggled with depression and anxiety, I owe my recovery to my pets.— online respondent. (Fraser and Taylor 2015–16)

> I suffer from a psychiatric illness. I am a dog lover and have a Labrador-cross called Pogo. He is my best friend and has been my rock and support for many years now. I cannot thank my loyal companion enough for the unconditional love and support he has provided me over the years. When I'm unwell, he stands by me and does not judge. The importance of friendship, love and companionship that my dog provides is beyond words.—online respondent. (Fraser and Taylor 2015–16)

The women who participated in our studies did not speak of the animals as second class or inferior to humans, but as family members that gave their lives meaning. A few participants noticed the potential for "animal lovers" to be dismissed and trivialized. Some women embraced being "gushy" or overly sentimental.

> I know I am lucky to have had our time together and can rely on what I know was a real bond to get me through the eye-rolling and eyebrow-lifting other non-animal-lovers give me when I gush about my bestest friend … Love you to infinity Nora-G Xxxoxxxo—online respondent. (Fraser and Taylor 2015–16)

> I'm like other animal crazies (as I fondly refer to them) in that I've been owned by several family pets, the most recent of which being my eleven-year-old dog, Sally—focus group respondent. (Fraser and Taylor 2015–16)

In both studies participants shared mostly happy, uplifting, and joyous images and stories about companion animals, perhaps reflecting the wider popular preoccupation with "pets," especially on social media. Yet, questions may be asked about the wisdom of emphasizing women's proximity to, and fondness for, domesticated animals. Eyebrows might also be raised at the apparent lack of critical content, particularly in relation to basic animal rights. No agreement was reached, or discussed at any length, about the love of animals needing to translate into political actions, such as the adoption of rescue animals, the support for animal sanctuaries, resistance to agribusiness and veganism. Without a critical analysis, studies such as ours might be read as exacerbating the potential exploitation and oppression of animals through the construction of them as "pets" for humans, reinscribing (even unwittingly) speciesist, dominant-submissive relationships. It is to this political analysis that we now turn, starting with the historical tendency to pair women with emotions and nervous maladies.

Women, anxiety, and animals

In contemporary society, anxiety is a problem for many whose health is affected by overwhelming, intrusive negative thoughts and feelings, and an impending sense of doom (Shekhar et al. 2005). Historically, women have been closely associated with nervous maladies and disparaged for being prone to hysteria and other nervous ailments (see, e.g., Bernheimer and Kahane 1990; Marcellus 2008; Weeks 2014). Today, anxiety is a problem that still disproportionately affects women. Current estimates of diagnosed anxiety disorders in the United States, for instance, show women outnumber men (2:1); however, it is difficult to determine whether this is due to more women than men presenting for treatment (Armstrong and Khawaja 2002).

The eight main types of medically diagnosed anxiety include (1) separation anxiety, or intense fears of being separated from love attachments; (2) generalized anxiety (disorder), which literally refers to anxiety generalized to all parts of life; (3) social phobia, or social anxiety (disorder), which refers to anxiety in social situations, especially anxiety about the likelihood of humiliation; (4) specific phobias, such as excessive fear of an animal, object, or process (such as air travel); (5) agoraphobia, such as fear of being in public places, leaving home and not having an easy exit if panic attacks occur; (6) post-traumatic stress (disorder), common to many trauma survivors, including those returning from war; (7) panic (disorder), often characterized by intense physical symptoms, such as dizziness and being short of breath; and (8) obsessive-compulsive disorder, involving repetitive rituals, such as frequent hand-washing (American Psychiatric Association 2017).

Medical diagnoses applied to mental health conditions, such as those described above, appear to be stable, certain, and universally applied classifications. These diagnoses can feel comforting and can help people make sense of experiences that can be overwhelming and intensely difficult to manage. For some, anxiety-related diagnoses provide a shorthanded way of expressing to the self and others, the mental health challenges being faced. For many anxiety sufferers, medical diagnoses and often accompanying prescription of medication influence, if not redefine, identities. What is so often forgotten is that both medical diagnoses, particularly those in mental health, and the experience of intense emotions (such as anxiety) are bounded by history, culture, and politics (see Conrad 2005). As with other emotions and psychological states (such as rage, terror, or catatonia), anxiety is not just an individual experience but also a collective, sociocultural one. Anxiety is not always experienced as or best understood to be a classifiable pathology. Nor is it always wise to "treat" it as if it is an individual experience, unrelated to social location (such as class, gender, ethnicity, and age) or the influence of the pharmaceutical industry, which, as Conrad argues, plays a major role with doctors constructing, defining, and prescribing drug treatment for a variety of anxiety-related conditions. As discussed below, rather than disaggregate anxiety from sociocultural contexts, we need to pay attention to living conditions productive of anxiety.

Anxiety and inequality

Anxiety is not confined to individual human psyches but intersects privilege and oppression (based on gender, class, race/ethnicity, age, and ability). Inequality produces anxiety, especially for oppressed groups experiencing discrimination and disadvantage. For many oppressed groups, social problems such as poverty, homelessness, and unemployment generate high levels of anxiety. To quote Haushofer and Fehr, "poverty correlates with unhappiness, depression, anxiety, and cortisol levels ... [and that] results from randomized controlled trials that show that cash transfers [to the poor] reduce distress and depression scores" (2014, 864). Goldsmith et al. (1996) noted the negative impact of unemployment on self-esteem, not just on anxiety and depression but also on the experience of alienation.

Rising social inequality and poverty rates for women (Fraser and Taylor 2016) and women's burden of disease risks associated with domestic violence, sexual assault, and legacy of child abuse (see Williams and Mickelson 2004) influence women's experiences of anxiety. Williams and Mickelson show that when women experience poverty and domestic violence, their resilience to anxiety is usually compromised. Stigma, shame, and victim blaming often exacerbate anxiety. Anxiety-producing gendered patterns of work—paid and unpaid— also play a role in the lives of many women, especially for women trapped in casual, underpaid work, performing the double shift and unpaid work in the family (Pease 2010).

Feminists have criticized the biomedical tendency to pathologize women for suffering anxiety, noting that contextual reasons for women's propensity for nervous ailments are usually ignored, "Gender differences between men and women in pay, social status, political power, burdens of domestic care and mothering, relationship inequalities and rates of domestic violence ... [and] or gendered ideals about parenthood and care responsibilities ... [explain] women's increased likelihood of experiencing mental health problems" (Malson and Nasser 2007, 10).

Gendered expectations of normative anxiety are evident, with women stereotypically portrayed as those who should worry and fret over others, through their assumed roles as wives, mothers, and homemakers. In other words, anxiety is not just the feelings, thoughts, and actions associated with worry, threats, fear, stress, and tension but also the codified expectations, or conventions, for socially sanctioned causes and expressions of anxiety.

High anxiety may distort experience, but it can also be a manifestation of life-threatening situations, such as seeking asylum (see Lindert et al. 2017). This is not a reference to the status anxiety of wealthy humans, but the gritty, collective angst of low-income groups trying to survive (Jetten et al. 2017). For some groups, especially oppressed groups, anxiety—or the impending sense of doom—is not just imagined but an eventuality.

Animals caught up in agribusiness are excellent (nonhuman) examples of structural oppression, and of how anxiety and an impending sense of doom becomes a reality, not just a fear. Irrespective of the position taken with regard to the legitimacy of agribusiness, the anxiety and terror experienced by "livestock" especially during slaughter can no longer be denied (Pachirat 2011). Dominant social and cultural conventions still converge through human carnistic defenses, justifying the incarceration of animals for reproductive products and use of body parts post-slaughter (Taylor and Fraser 2017). However, a large body of research is now available showing the negative impact of forced reproduction and milking for cows, sheep, goats, and camels, and the analogous ways in which lactating women and other animals may be commodified (see Cohen 2015), and the sexual politics of meat (see Wright 2015, referencing Carol Adams).

Abundant feminist critiques are available of the historical inclination to associate emotion with women and reason with men, divorcing emotion from reason and tying emotion to supposedly fragile, brittle forms of femininity (Jaggar and Bordo 1989). Feminists have documented how discourses of madness have been used to control women, just as they have pointed out that women's perceived closeness to nature that has led to their being designated emotional (cf. rational) and in need of seclusion from the public sphere, or control while operating in it (Marcellus 2008). As Malson makes clear, this is often achieved through the animalization of women, particularly focusing on reproduction which is, arguably, the site of the most control of the female body: "The 'maternal body', like 'the body' of dualist discourse, is constructed as the antithesis of the mind, will or spirit. It is animal-like—cow-like—uncontrolled, and excessive" (1997, 237). Those women who fail or refuse to reproduce are still liable to be constructed as lesser women (Letherby 2002).

"Old maids" and "crazy cat ladies"

Single women and women who choose not to have children have long been viewed with suspicion, especially if their single status spans more than a few years (Fraser 2003). To quote Gordon, "The old maid stereotype is related to a perception of single women as lacking something, being incomplete, deviating from the norm and the normal" (1994, 129). Graham and Rich describe the pro-natalist discourses still in circulation in Australian media—media that produces an array of messages that pity childless women and reprimand those who decide not to have children (2012). Bay-Cheng and Goodkind note that "Conventional gender roles and rules for young women center on their relationships to men" but add that these conventions interface with neoliberal imperatives on different socio-demographic groups of women, toward "self-focused striving," exemplified by staying single so as to "stay on track" and "get ahead" (in education or work terms) (2016, 193). Blackstone describes how increasing numbers of Western women are creating families that do not include children, noting that they are often overlooked or treated with ambivalence (2014). Focusing on human–dog relationships in multispecies households, Charles refers to them as post-human families (2016), as does Irvine and Laurent (2017). However, several heteronormative stereotypes still apply to single women, especially as they age and if they live in close company with animals (McKeithen 2017).

Throughout Western history this denigration of, and suspicion about, women who have close relationships with other species has held ground. For instance, at the turn of the twentieth century, and with the rise of women's antivivisectionism, American neurologist Charles Loomis Dana created a new medical condition in 1909 called "zoophilpsychosis," which referred to those with a heightened concern for animals as afflicted with a mental illness (Buettinger 1993, 277). To some extent this notion lives on. Take, for instance, the "crazy cat lady" that implies there's something wrong with women who love cats "too much" (McKeithen 2017, 127). Stereotypically, "crazy cat ladies" are those women who place the needs of cats above humans, a move that is to be scorned and mocked. Some so-called "crazy cat ladies" keep multiple cats, sometimes to the point of hoarding. "Hoarding can be thought of as pathological over-attachment to animals, and has been considered to be a manifestation of several forms of psychopathology" (Herzog 2007, 11). Hoarding animals can be a serious problem for companion animals, especially if they are overcrowded, underfed, or otherwise neglected. However, questions need to be asked about who gets to attribute such labels (see Probyn-Rapsey chapter in this volume).

The kind of reductive thinking that assumes women who have empathy for other animals are somehow "faulty," and/or that their inability to maintain productive (in both senses of the word) relationships with other humans closes down alternate ways of seeing, experiencing, and thinking about the world. In terms of our relationships with other animals, it ensures that emotion and empathy are relegated to "unscientific" ideas about other animals and rationalist critiques of anthropomorphism (Gruen 2015). As Davies points out, "the language of Western science—the reigning construct of male hegemony—precludes the ability to express the experiential realities it talks about" (1995, 208).

A feminist animal studies

We are not alone in our argument that a distinctly feminist animal studies is needed (see, e.g., the special edition of *Hypatia*, Gruen and Weil 2012). With Gruen and Weil, we are in agreement:

One clear commonality is the need to maintain feminist, ethical, and political commitments within animal studies—commitments to reflexivity, responsibility, engagement with the experiences of other animals, and sensitivity to the intersectional contexts in which we encounter them. Such commitments are at the core of a second, related area of concern, that of the relationship between theory and practice. Animal bodies, we can all agree, must not be "absent referents" in animal studies. (2012, 493)

Feminism also needs animal studies. Mohanty (2003) refers to "feminism without borders" and while she stops short of acknowledging species as a border, her work can be fruitfully extended:

Feminism without borders ... acknowledges the fault lines, conflicts, differences, fears, and containment that borders represent. It acknowledges that there is no one sense of a border, that the lines between and through nations, races, classes, sexualities, religions, and disabilities [and species] are real—and that a feminism without borders must envision change and social justice work across these lines of demarcation and division. I want to speak of feminism without silences and exclusions in order to draw attention to the tension between the simultaneous plurality and narrowness of borders and the emancipatory potential of crossing through, with, and over these borders in our everyday lives. (2)

In *Speaking Up for Animals*, Kemmerer puts it plainly, "Women who seek equality must not support the oppression of nonhuman animals ... if women are ever to achieve equality we must topple hierarchy in total" (2012, 5). Maintaining a commitment to anti-oppressive practice is difficult, if not impossible, without recognizing species oppression, given all oppressions share similar structural expression and mechanisms. As Jenkins points out, feminists need to take species seriously because:

[F]eminism is perhaps best positioned to take on questions of the animal. This is manifest in feminist theory's commitment to the materiality of the body, to attending to those bodies most vulnerable to abuse, to exposing the logic of exclusion and the politics of abjection, and perhaps most important ... In its commitment to thinking about, and critiquing its own participation in, the ethics of representation and "speaking for" others. (2012, 517)

Brown (2016) explains that liberation discourses must not separate the oppression of animals from humans, noting that domestication has affected many women and animals negatively, with both groups liable to be cast as mindless and rightful subjects of men's servitude.

Feminism and animal studies share the injunction to praxis. A key part of praxis is alliance building. A feminist approach to alliance building recognizes similar structures of oppression but allows room for difference. As Gruen points out, "The feminist care tradition in animal ethics urges us to attend to other animals in all of their difference including differences of power within systems of human dominance in which other animals are seen and used as resources or tools" (2015, 36).

Useful ideas come from feminist work on forming and maintaining alliances across difference. This work stresses, as one might expect from feminist work, that the personal is the political and turns on recognizing the similarity behind different structures of oppression. To paraphrase Ann Bishop (2002), we cannot try to end our oppression by ignoring the oppression of others, or subordinating others, as each form of oppression interacts in a self-perpetuating system of domination. In turn, a feminist critique of individual competition is also required, with attention given to how competition can unhelpfully divide us and position some as superior and others as inferior, "The whole thing rests on a world-view that says we must constantly strive to be better than someone else. Competition assumes that we are separate beings—separate from each other, from other species, from the earth" (Bishop 2002, 19).

This chapter stands as a corrective to this kind of hierarchical thinking by considering how women express emotional connections to other animals, and by doing so in a non-pathologizing way that takes the interspecies relationships considered as "normal," "functional," and positive. The very act of allowing those expressions and seeing this allowance itself is a political act: "establishing the legitimacy of outcast experiences is precisely the political cultural work that needs to be carried out in real life for the sake of all beings disenfranchised by sanctioned value systems" (Scholtmeijer 1995, 231).

Caregiving across species

When we speak of care we are referring to both caring for and caring about others (Fraser et al. 2018). To qualify as care, it needs to be undertaken with kindness and without duress. Far from being easy, mundane, or routine, "quality" care is personalized and thoughtful (Fraser et al. 2018). Feminist sociologist Arlie Hochschild (1979, 1990, 2002) helped us understand the work and complexities involved in caring, especially the emotional labor. Hochschild defines emotion work as "the act of trying to change in degree or quality an emotion or feeling ... it refers to the effort—the act of trying—and not to the outcome, which may or may not be successful ... [it differs from] emotion 'control' or 'suppression'" (1979, 561).

Hochschild (1979, 1990, 2002) showed how emotions, not just behaviors, are subject to social rules productive of expectations of how emotional management should be undertaken and by whom. This is particularly relevant to women and companion animals in private domestic settings, where caregiving is often taken for granted or assumed to reflect their inherent nurturing capacities.

When speaking of connections to "their" animals, participants of the projects, *In Good Company* and *Whatisitaboutanimals*, often expressed taken-for-granted notions of caring relationships, based on reciprocity, particularly mutual understanding (also see Donovan [1990] 1996). The reciprocal nature of their relationships was a key component to seeing their animal companions as part of the family: as agentic, interactive, and emotionally attuned beings.

> I think she can understand me ... If I'm you know anxious or anything, she will stay really close to me and if I'm a little bit sad, she will kind of just look at me and come up to me and she thinks she is a lapdog, but she is 23 kilos and she gets her whole body like onto me up here and she knows what is going on.—focus group respondent. (Fraser and Taylor 2017)

Many women in our studies noticed the significant forms of emotional labor performed by their companion animals, evident in expressions such as "the dog would know I was always crying and she would come up and make sure I was OK" (focus group respondent in Fraser and Taylor 2017). Many of the animals were understood by participants to have offered themselves as therapists, nurses, and confidantes:

> He [dog] slept for hours and hours ... and this little puppy actually wedged herself at the top of his pillow just on the top of his head against the bed and just stayed there. She didn't have to get up or anything. She just knew he wasn't right or she thought he needed her somehow and she just stayed there and it was amazing.—focus group respondent. (Fraser and Taylor 2017)

Several participants expressed how important it was to them to care for abused and neglected animals, and in one instance spelt out how much pleasure could come from watching empathy shown across species, from one abused or abandoned animal to another:

> I love this face [dog]. I can't get enough of it! Stussy is the naughtiest of dogs but also the most giving. She knows when I am sick or sad. She knows when I'm cold and need a furry spoon. She has been an amazing foster sibling and done so much to help the damaged, abused and neglected dogs who have come in to our care. It's easy to forgive her for all the doonas [bed quilt] and pillows she has eaten.— online respondent. (Fraser and Taylor 2015–16)

Interspecies care, alliance building, and solidarity

Feminist animal studies refuse to pathologize women's deep emotional connections with other animals but instead makes room for an empathic understanding of them. From this perspective, the work that animals perform for humans is noticed and valued, such as attending to emotional needs, soothing anxiety, and providing support that can be therapeutic and healing. Recognizing the care work animals do for humans is crucial if we are to avoid simply taking this work for granted and/or to avoid positing them as tools whose main or only purpose is to help improve human well-being. As Coulter notes, when an animal-centric approach is taken, one that refuses the idea of animals as objects and/or as subordinate to humans, we "see" stories of animal–human cooperation and relationships differently: "The local-global evidence about animals' lives contains inspiring and novel examples of … how people and animals work together with respect and kindness, of animals' diverse and even surprising contributions, and of laudable, moving expressions of compassion within and across species" (2016, 139–140).

Interspecies alliances are tied to the recognition of respectful, mutually beneficial and life-sustaining relationships that can occur across species, human and otherwise (also see von Essen and Allen 2017). Admittedly, becoming allies with other animals is complicated by species differences, particularly communication styles. However, there are other powerful, non-deliberative (not talk-based) ways of building alliances, for instance through the preparedness of allies to perform emotional labor with each other. The challenge is to notice the work that allies do to sustain good relations.

While the broader concept of alliances is underdeveloped in human–animal studies literature, initial searches of mainstream scientific studies show mostly primate studies, apes and monkeys as the subjects (e.g., Cheney and Seyfarth 1986; Harcourt and de Waal 1992), although some bottle-nosed dolphin studies also show intragroup alliances (e.g., Connor et al. 1999). In these accounts, alliances are built on collaboration and teamwork, for many reasons, such as mating, protecting each other, solving problems, play and socialization, food finding, and the rearing of young (Cheney and Seyfarth 1986; Harcourt and de Waal 1992; Connor et al. 1999). However, it should be noted that similar to the concept of "animal lovers," interspecies alliances do not escape the problem of selecting some species, or representatives from selected species to love, without regard for the needs and rights of other species or concern over their maltreatment—for instance, loving dogs but eating cows and drinking their milk. This is why we are emphasizing the need for alliances of both care and solidarity.

According to Bayertz (1999, 3), solidarity "today stands, largely unexplained in relation to complementary terms such as 'community spirit' or 'mutual attachment', 'social cooperation' or 'charity', and—from time to time—'brotherly love' or 'love of mankind'. However, we need to think about solidarity beyond men and humans." To quote Rock and Degeling (2015, 4), " more-than-human solidarity builds on two existing concepts: 'environmental health justice' and 'multi-species flourishing.'" Herzog (2017, 3) reminds us that "People who felt high solidarity with animals were more concerned with animal welfare and more apt to think of themselves as part of the natural world."

Building on the idea of alliances with other species, versus the ownership of them, Coulter proposes "interspecies solidarity," which she describes as "an idea, a goal, a process, an ethical commitment, and a political project" (2016, 3). In keeping with her feminist analysis, she notes that key ideas from the feminist ethic of care for other animals are necessary to the idea of interspecies solidarity: "The concept of solidarity is underscored by ideas of empathy" (150). Key to this idea of caring is that "individual acts of solidarity matter, and they can disrupt dominant perceptions and power relations. They can also set a domino effect in motion which propels a broader set of processes. Moreover, solidarity can prompt and inform larger, collective forms of political work. Caring can be and can become political" (152).

Perhaps it is interspecies alliances of care and solidarity that some of the participants meant through their images, meanings that we did not pick up at first. Consider, for instance, the submission of a photograph online entitled "Me and Dizzy," which depicts a close-up image of a paw in a handshake with a human. An excerpt of the accompanying text submitted with this image included, "Dizzy [dog] has been that confidante, who always calmly listens and offers a comforting, thoughtful and understanding. Unspoken. But nonetheless, shared" (Fraser and Taylor 2015–16).

Our position is that if interspecies care and empathy is shown, the life of a companion animal does not have to be inevitably bounded by servitude. Interspecies alliances of care and solidarity are possible, across and in between humans and companion animals (Rock and Degeling 2015). However, there is a caveat: to meet our criteria for alliances of care and solidarity, there cannot be relationships of human domination and animal subordination. Strict hierarchical arrangements are an anathema to the alliances of care and solidarity. In strict hierarchical arrangements, generosity may be expressed but only through philanthropy and benevolence. It is not a question of allies needing to be the same in rank or similar to each other in personal characteristics. Diversity is, and has been, crucial to alliance building, across borders, classifications, and social movements (Bishop 2002). Interspecies alliances are inherently diverse and must operate with regard to all parties' health and well-being.

The intensely personal experience of anxiety should not be allowed to eclipse the sociocultural contexts that give rise to anxiety in the first place. Biomedical and neoliberal approaches to health, including experiences of anxiety, are approached with caution for their tendency to pathologize and hyper-individualize experiences, especially women's experiences. Because anxiety shapes the way the self and others are experienced culturally, socially, and politically, it is crucial to recognize the interplay of anxiety and gender politics.

From a feminist animal studies perspective, women—including women reporting high levels of anxiety—are believed and respected, rather than teased or mocked, for forming close emotional ties with animals. We have argued for feminism and animal studies to work in simpatico, rather than against each other, to allow for the many cross-species connections in experiences of oppression and to open up possible alliances. Our aim has been to dignify interspecies relationships and recognize the valuable work that companion animals perform for humans. This includes emotional labor, such as soothing anxiety, which for both women and other animals cannot be disaggregated from social, cultural, and political contexts.

References

American Psychiatric Association (2017), "What Are Anxiety Disorders?" Available online: https://www.psychiatry.org/patients-families/anxiety-disorders/what-are-anxiety-disorders (accessed February 22, 2018).

Armstrong, K. A. and Khawaja, Nigar G. (2002), "Gender Differences in Anxiety: An Investigation of the Feb Symptoms, Cognitions, and Sensitivity towards Anxiety in a Nonclinical Population," *Behavioural and Cognitive Psychotherapy* 30: 227–231.

Bay-Cheng, L. Y. and Goodkind, S. A. (2016), "Sex and the Single (Neoliberal) Girl: Perspectives on Being Single among Socioeconomically Diverse Young Women," *Sex Roles* 74 (5–6): 181–194.

Bayertz, K. (1999), "Four Uses of 'Solidarity,'" in K. Bayertz (ed.), *Solidarity*, 3–28, Dordrecht: Springer.

Bernheimer, C. and Kahane, C. (1990), *In Dora's Case: Freud—Hysteria—Feminism*, New York: Columbia University Press.

Bishop, A. (2002), *Becoming an Ally: Breaking the Cycle of Oppression in People*, 2nd edition, New York: Zed Books.

Blackstone, A. (2014), "Doing Family without Having Kids," *Sociology Compass* 8 (1): 52–62.

Brown, K. (2016), "A Feminist Analysis of Human and Animal Oppression: Intersectionality among Species" [Departmental Presentation]. Available online: http://cedar.wwu.edu/scholwk/Departmental_Presentations/Women_Gender_Sexuality_Studies/2/ (accessed February 22, 2018).

Buettinger, C. (1993), "Antivivisection and the Charge of Zoophil-Psychosis in the Early Twentieth Century," *Historian* 55 (2): 277–288.

Charles, N. (2016), "Post-Human Families? Dog-Human Relations in the Domestic Sphere," *Sociological Research Online* 21 (3): 1–12.

Cheney, D. L. and Seyfarth, R. M. (1986), "The Recognition of Social Alliances by Vervet Monkeys," *Animal Behaviour* 34 (6): 1722–1731.

Cohen, M. (2015), "Regulating Milk: Women and Cows in France and the United States," *American Journal of Comparative Law* 65: 469–526.

Connor, R. C., Heithaus, M. R., and Barre, L. M. (1999), "Superalliance of Bottlenose Dolphins," *Nature* 397 (6720): 571.

Conrad, P. (2005), "The Shifting Engines of Medicalization," *Journal of Health and Social Behavior* 46 (1): 3–14.

Coulter, K. (2016), *Animals, Work and the Promise of Interspecies Solidarity*, London: Palgrave.

Davies, K. (1995), "Thinking like a Chicken: Farm Animals and the Feminine Connection," in C. Adams and J. Donovan (eds.), *Animals and Women: Feminist Theoretical Explorations*, 192–212, Durham and London: Duke University Press.

Donovan, J. ([1990] 1996), "Animal Rights and Feminist Theory," in J. Donovan and C. Adams (eds.), *Beyond Animal Rights: A Feminist Caring Ethic for the Treatment of Animals*, 34–59, New York: Continuum.

Fraser, H. (2003), "Narrating Love and Abuse in Intimate Relationships," *British Journal of Social Work* 3: 273–290.

Fraser, H. and Taylor, N. (2015–16), "What Is It about Animals?" Available online: www.whatisitaboutanimals.com (accessed September 27, 2017).

Fraser, H. and Taylor, N. (2016), *Neoliberalization, Universities and the Public Intellectual: Species, Gender and Class and the Production of Knowledge*, London: Palgrave.

Fraser, H. and Taylor, N. (2017), "In Good Company: Women, Companion Animals and Social Work," *Society & Animals* 25 (4): 341–361.

Fraser, H., Taylor, N., and Morley, M. (2018), "An Intersectional Perspective on Ethics, Principles and Practices," in B. Pease, A. Vreugdenhil and S. Stanford (eds.), *Critical Ethics of Care in Social Work: Transforming the Politics and Practices of Caring*, 229–240, London: Routledge.

Goldsmith, A. H., Veum, J. R., and William, D. (1996), "The Impact of Labor Force History on Self-Esteem and Its Component Parts, Anxiety, Alienation and Depression," *Journal of Economic Psychology* 17 (2): 183–220.

Gordon, T. (1994), *Single Women: On the Margins?* New York: New York University Press.

Graham, M. and Rich, S. (2014), "Representations of Childless Women in the Australian Print Media," *Feminist Media Studies* 14 (3): 500–518.

Gruen, L. (2015), *Entangled Empathy: An Alternative Ethic for Our Relationships with Other Animals*. New York: Lantern Books.

Gruen, L. and Weil, K. (2012), "Invited Symposium: Feminists Encountering Animals," *Hypatia* 27 (3): 492–526.

Harcourt, A. H., and De Waal F. B. M. (eds.) (1992), *Coalitions and Alliances in Humans and Other Animals*, Oxford: Oxford University Press.

Haushofer, J. and Fehr, E. (2014), "On the Psychology of Poverty," *Science* 344 (6186): 862–867.

Herzog, H. (2007), "Gender Differences in Human–Animal Interactions: A Review," *Anthrozoös* 20 (1): 7–21.

Herzog, H. (2017), "How Much Solidarity Do You Feel with Animals?" *The Animal Studies Repository*, 1–5. Available online: http://animalstudiesrepository.org/aniubpos/33/ (accessed February 22, 2018).

Hochschild, A. R. (1979), "Emotion Work, Feeling Rules, and Social Structure," *American Journal of Sociology* 85 (3): 551–575.

Hochschild, A. R. (1990), "Ideology and Emotion Management: A Perspective and Path for Future Research," in T. D. Kemper (ed.), *Research Agendas in the Sociology of Emotions*, 117–142, Albany, NY: SUNY Press.

Hochschild, A. R. (2002), "'Emotional Labour,'" in S. Jackson and S. Scott (eds.), *Gender: A Sociological Reader*, 192–196. London: Routledge.

Irvine, L. and Laurent, C. (2017), "More-than-Human Families: Pets, People, and Practices in Multispecies Households," *Sociology Compass* 11 (2). Available online:

http://onlinelibrary.wiley.com/doi/10.1111/soc4.12455/full (accessed February 22, 2018).

Jaggar, A. M. and Bordo, S. (eds.) (1989), *Gender/Body/Knowledge: Feminist Reconstructions of Being and Knowing*, New Brunswick, Rutgers University Press.

Jenkins, S. (2012), "Returning the Ethical and Political to Animal Studies," *Hypatia* 27 (3): 504–510.

Jetten, J., Mols, F., Healy, N., and Spears, R. (2017), "'Fear of Falling': Economic Instability Enhances Collective Angst among Societies' Wealthy Class," *Journal of Social Issues* 73 (1): 61–79.

Kemmerer, L. (ed.) (2012), *Speaking Up for Animals: An Anthology of Women's Voices*. London: Routledge.

Letherby, G. (2002), "Childless and Bereft? Stereotypes and Realities in Relation to 'Voluntary' and 'Involuntary' Childlessness and Womanhood," *Sociological Inquiry* 72 (1): 7–20.

Lindert, J., Von Ehrenstein, O. S., Wehrwein, A., Brahler, E., and Schäfer, I. (2017), "Anxiety, Depression and Posttraumatic Stress Disorder in Refugees—A Systematic Review," *Psychotherapie, Psychosomatik, Medizinische Psychologie* 68 (1): 22–29.

Malson, H. (1997), "Anorexic Bodies and the Discursive Production of Feminine Excess," in J. Ussher (ed.), *Body Talk: The Material and Discursive Regulation of Sexuality and Madness and Reproduction*, 223–245, London; New York. Routledge.

Malson, H. and Nasser, M. (2007), "At Risk by Reason of Gender," in M. Nasser, K. Baistow, and J. Treasure (eds.), *The Female Body in Mind: The Interface between the Female Body and Mental Health*, 3–16, London: Routledge.

Marcellus, J. (2008), "Nervous Women and Noble Savages: The Romanticized 'Other' in Nineteenth-Century US Patent Medicine Advertising," *The Journal of Popular Culture* 41 (5): 784–808.

McKeithen, W. (2017), "Queer Ecologies of Home: Heteronormativity, Speciesism, and the Strange Intimacies of Crazy Cat Ladies," *Gender, Place & Culture* 24 (1): 122–134.

Mohanty, C. (2003), *Feminism without Borders: Decolonising Theory, Practicing Solidarity*, Durham and London: Duke University Press.

Pachirat, T. (2011), *Every Twelve Seconds: Industrialized Slaughter and the Politics of Sight*, New Haven, CT: Yale University Press.

Pease, B. (2010), *Undoing Privilege: Unearned Advantage in a Divided World*, London: Zed Books.

Rock, M. J. and Degeling, C. (2015), "Public Health Ethics and More-than-Human Solidarity," *Social Science & Medicine* 129: 61–67.

Scholtmeijer, M. (1995), "The Power of Otherness: Animals in Women's Fiction," in C. Adams and J. Donovan (eds.), *Animals and Women: Feminist Theoretical Explorations*, 231–262, Durham and London: Duke University Press.

Shekhar, A., Truitt, T., Rainnie, D., and Sajdyk, R. (2005), "Role of Stress, Corticotrophin Releasing Factor (CRF) and Amygdala Plasticity in Chronic Anxiety," *Stress* 8 (4): 209–219.

Taylor, N. and Fraser, H. (2017), "Slaughterhouses: The Language of Life, the Discourse of Death," in J. Maher, H. Pierpoint, and P. Beirne (eds.), *The Palgrave International Handbook of Animal Abuse Studies*, 179–199, London: Palgrave.

Taylor, N., Fraser, H., and Riggs, D. W. (2017), "Domestic Violence and Companion Animals in the Context of LGBT People's Relationships," *Sexualities*, 1–15. Available online: http://journals.sagepub.com/doi/full/10.1177/1363460716681476 (accessed March 15, 2018).

Von Essen, E., and Allen, M. P. (2017), "Solidarity between Human and Non-Human Animals: Representing Animal Voices in Policy Deliberations," *Environmental Communication* 11 (5): 641–653.

Weeks, J. (2014), *Sex, Politics and Society: The Regulations of Sexuality since 1800*, London: Routledge.

Williams, S. L. and Mickelson, K. D. (2004), "The Nexus of Domestic Violence and Poverty: Resilience in Women's Anxiety," *Violence against Women* 10 (3): 283–293.

Wright, L. (2015), *The Vegan Studies Project: Food, Animals, and Gender in the Age of Terror*, Georgia: University of Georgia Press.

Part III

DYSFUNCTION

Chapter 11

THE "CRAZY CAT LADY"

Fiona Probyn-Rapsey

If you ever work on a project about "crazy cat ladies," you can expect some smiles of the wry, knowing kind, followed up by a flurry of "crazy cat lady" (CCL) memes, mugs, socks, and links to various CCL products. Many of them are pretty funny, especially the picture of the box of kittens labeled "Crazy cat lady start up kit" and the socks emblazoned with "you say 'crazy cat lady' like it's a bad thing." These jokes attest to a level of sympathy for and fascination with crazy cat ladies within popular culture. I've long been fascinated by the "crazy cat lady" as a cultural trope, a sort of "folk devil" whose appearance plays on broader anxieties attached to femininity and animality. Putting the memes, socks, mugs, pyjamas, fridge magnets aside for a moment, this chapter takes a more critical look at the CCL. I explore the current popularity of the CCL in three connected ways. Firstly, as a *gendered* cultural trope that is mobilized in both negative and positive ways to exemplify the feminization of concern for human–animal relations. Secondly, I examine how the CCL gets tangled up with the animal hoarder: someone who "hoards" or collects animals and keeps them as their self-declared "rescuer," often to protect them from some other terrible fate (neglect and cruelty) that then becomes realized in her own hands. The research on animal hoarding is fascinating in this regard, because it essentially plays chicken and egg with the "crazy cat lady," replicating gender stereotypes in its discussion of the disorder it attempts to outline. While animal hoarding literature situates the CCL as a dangerous obstacle to proper diagnosis and understanding of animal hoarding cases, I then take this idea one step further and discuss whether or not the CCL might be not just a "cute face" of the animal hoarder but also the "folk devil" for the industrial scale hoarding of animals that persists in factory farming situations. The three elements—"crazy cat lady," animal hoarder, and factory farmer—are connected, I suggest, by a broader phenomenon of the intensification of animal keeping in Western modernity, a period in which animals are simultaneously more numerous, less visible but more intensively "kept" (Harrison 1964; Vialles 1994; O'Sullivan 2015; Pachirat 2015). In this chapter, I'll pull on the thread of the crazy cat lady trope and see how she leads us to industrialized hoarding in the form of factory farming.

A popular history of the CCL

When Eleanor Abernathy arrived on the scenes of *The Simpsons* in 1998, the madness of her rants and raves, accompanied by cats, made her perhaps the most recognizable and transportable first "crazy cat lady" of popular culture. She pops up from time to time throughout the show and reveals in one episode that she has a law degree from Yale and a medical degree from Harvard, a highly accomplished "mad" woman. As a figure of derision, and of dysfunctional and misdirected talent, she exemplifies the fine line between the CCL and the animal hoarder, the latter being the former's much less "funny" companion. Also walking a fine line between humor and loathing is J. K. Rowling's character Dolores Jane Umbridge, whose militancy, empathy failures, and dictatorial style while headmistress of Hogwarts goes hand in hand with her pink outfits, girlish giggle, and her vast collection of kittens framed on her office walls (2003). As a "crazy cat lady" (without any actual cats), Umbridge is also a figure of gendered dysfunction whose "love" for cats is indicative of her contempt for the "real" lives of others (see also McKeithen 2017). Going back to 1975, the documentary film *Grey Gardens* follows the decline of an American aristocratic mother and daughter and shows the Beales' living conditions shifting from ancestral privilege (they are related to Jaqueline Bouvier Onassis) to one of squalor, where the mother and daughter share their once grand house with raccoons and many cats whom they claim to breed. Their exposure within the film parallels that of other animal hoarding documentaries, where the hoarder's life is exposed by camera and crew, and we, the audience, asked implicitly to recoil and then reinscribe a "line" between animal and human, cleanliness and squalor, civility and barbarism, domestic and feral (Krasner 2017).

While we might situate these popular culture CCLs as a modern and also Western phenomenon, the association between single (often older) women living in the company of cats goes much further back into the archives: women as witches with their "familiars," unhinged from human society, living at odds with the rest of the community and seeking instead the company of animal Others. The CCL gains mobility and meaning because it is a reminder of the historical, cultural association between mature women, animals, and the irrational; that not only are our relations with animals sentimental (read "trivial") but they are also made possible by having a shared status of mindless irrationality and animality. As Adams and Donovan show in their introduction to *Animals and Women* (1995), women and particularly feminists have historically put a great deal of effort into attempting to show that women are *not* animals. "Crazy cat ladies" illustrate just why that might be the case, as well as the importance of reclaiming the association in different, more positive terms.

In a recent positive review of the mainstreaming of the CCL, Linda Rodriguez McRobbie suggests that the "crazy cat lady" is actually losing its potency as a stigma and she puts this down to three things. One is that women no longer feel threatened by the spinster image that women with cats seems to conjure up (marriage is on the decline). Secondly, she suggests that pet culture and capitalism encourages grand displays of affection for animals, and that, thirdly, love for cats is made visible and therefore more "normal" by social media. The article also highlights how celebrities (like Taylor Swift) are taking up the moniker of CCL, changing the image of the CCL from Eleanor Abernathy, who throws cats at people, to someone who seeks out photo-opportunities with them. No longer, "crazy," McRobbie's story concludes that the "cat lady" is now "chic, she's young, she's got a good job, she's not always a lady, and it doesn't matter if she's not married … being a cat lady is no longer so taboo" (2017).

This shift away from the "taboo" is also seen most readily in feminist animal studies. For example, in her prescient and insightful essay "Bitch, Bitch, Bitch: Personal Criticism, Feminist Theory, and Dog Writing," Susan McHugh suggests that women who elect to "share their lives with female canines … become targets of criticism as 'indulgent' for focusing on their dogs" (2012, 616). But she goes on to say that women who take the risk are rewarded by the fact that becoming "down with the bitches … offers a way of recalibrating centers beyond the abstract model of the lone 'authoritative' human individual, and thereby of rethinking feminist politics as intra-active, even trans-species, from the ground up" (632). This may well be what is at the heart of the embrace of the CCL image too: a desire to rethink feminist politics in the company of dogs/cats/other animals. But not too far from this desire is also the backlash that McHugh signals, or an undercurrent of sexism and animalization that situates women and animals as forming a dangerous alliance: dangerous to her precarious grip on humanness and its psychic corollaries (reason, rationality, transcendence). At the heart of the "joke" about the CCL is that she has gone too far, "gone to the dogs"/cats, that her love of cats is fundamentally misanthropic.

It's important to bear in mind that when it comes to the animals that she might be "keeping," none of this is particularly funny either for her or for the animals who might have to be rescued from their "rescuers." CCL as a cultural archetype is implicated in animal hoarding—where the "cat lady" becomes the "crazy cat lady" and her care becomes toxic for the animals she "rescues." Just what that toxic care looks like for the animals is frequently overlooked, to the detriment of the animals who suffer profoundly, and are often euthanized as a result (see Pollak et al. 2014; Morrow et al. 2016; and Joffe et al. 2014). According to the literature on animal hoarding, it is predominantly women who are responsible for these cruelty cases, leading more than one expert to comment that the "crazy cat lady" is indeed a stereotype and yet "there is some truth to it" (Arluke and Killeen 2009, 167). However, this is where we need a more complex picture of how gender plays its part in the discussion of animal hoarding. Given her apparent ubiquity in popular culture, it is something of a surprise to find that the evidence for the crazy cat lady (as animal hoarder) is still rather patchy.

Animal hoarding studies

American sociologists have been studying animal hoarding since the late 1990s. A leading expert on animal hoarding is US sociologist Gary Patronek, who in 1999 published a study that concluded that most hoarders were women (76 percent), aged sixty or over (46 percent), single, and living alone. This early study is at the center of multiple studies that follow, and its statistics are cited repeatedly, but often without Patronek's own original equivocations or caveats about the reliability and accuracy of the data. Patronek's study was based on fifty-four case reports sent to him from ten animal control agencies and humane societies across the United States. Three problems with the original mapping of animal hoarders emerge. The data set is very small, and yet the reporting in terms of percentages gets repeated in a way that provides an illusion of bulk. The source for the data (animal control agencies) is also not without problems. It is difficult for Patronek's study to account for the possibility of interpretive bias—an expectation from agents, neighbors, or councilors—that women with multiple pets are deranged and deviant and more likely to be suspected of, and ultimately reported for, hoarding-related offenses. It seems that men do not carry the same cultural baggage when it comes to "keeping" animals. We get a hint of this gender bias in favor of reporting women for hoarding when Patronek notes that agencies found it difficult to prosecute men for animal hoarding offenses because the men did not fit the image of the hoarder, an image that precedes the ascendancy of the condition or crime. If a judge can't see a man as an animal hoarder, then it's likely that neighbors will also overlook them; they are, after all, sharing the same cultural landscape that depicts women and men as having specifically gendered relations to animals and animality. Although studies since then stress that animal hoarding stretches across multiple demographics, they keep returning to this idea that most animal hoarders are female, without all the caveats in place. Consequently, the prevalence of women as animal hoarders gets repeated uncritically throughout the literature.

Most research on the prevalence and demographics of animal hoarders assert an awareness of the stereotype of the cat lady and admit that it may skew reporting, while also suggesting, rather unconvincingly, that the facts remain somehow untouched by the influence of that very stereotype. In other words, the literature on animal hoarding seems to want to have it both ways, which is precisely how stereotypes operate. As literary theorist Homi Bhabha writes, a stereotype "is a form of knowledge and identification that vacillates between what is always 'in place', already known, and something that must be anxiously repeated" (1994, 66). It is this "anxious repeating" that signals its precarious relationship to reality:

> Animal hoarders come from varied backgrounds, despite the stereotype of
> the neighbourhood "cat lady" who is an older, single female, living alone. As
> with many stereotypes, there is an element of truth to this image. In one study

(Worth and Beck 1991) just 70% of the sample were unmarried women who had cats, while in another study (Patronek 1999), 76% of the sample were women, 46% were over sixty years of age, most were single, divorced, widowed and cats were most commonly involved. (Arluke and Killeen 2009, 167)

Urban legends tend to be fairly resistant to "objective" fact-finding. That is, they can work to structure reality while also being self-reflexively repudiated. In the case of women who are animal hoarders, this ambivalence about how to "read" her activities persists. One example of this can be seen in Svanberg and Arluke's reading of the "Swan Lady," which recounts the case of a Swedish woman, convicted for animal cruelty, who rescued and kept a large number of swans in her apartment. The authors argue that there was confusion in the media about whether to classify the offender as an animal lover or a criminal. Svanberg and Arluke suggest that instead "the Swan Lady became a kind of heroine, or rather anti-heroine, at least according to some—not because she had rescued swans per se but because she took these birds into her apartment … the epithet 'Swan Lady' added to the media's spin on this case as a charming, albeit ill-informed, behaviour" (2016, 65). They point out that while there was no fitting term for animal hoarding in Sweden, the community would have been familiar with the "crazy cat lady" through American television, especially Eleanor Abernathy from *The Simpsons*. This encouraged a view of the woman's activities as more akin to an "absurd anti-heroine" than a case of animal cruelty. Their reading suggests the global reach of such a stereotype (via *The Simpsons*) and its capacity not only to move between borders and languages but also to influence, construct, and reinforce the social norms it appeals to. For the authors, the emphasis on the "Swan Lady" as "absurd anti-heroine" served to "excuse" the cruelty that was felt by the animals she claimed to care for and exemplified how the reporting of such cases is very often anthropocentric. It is also an example of something pointed to earlier, which is that the "Swan Lady," like the CCL or woman who is an animal hoarder is likely to be partly "excused" on the basis of gender (it is *merely* caring gone wrong). But she is also likely to be *singled out* because of her gender.

This ambiguity is perhaps what makes the CCL and woman animal hoarder such an attractive media story. The framing of animal hoarding cases is often not just anthropocentric but also misogynistic. When misogyny combines with anthropocentrism it works to scapegoat women and also confine the issue to the "madness" of women, that is, to a symptom of gender rather than a broader issue of human arrogance. This effectively brackets a broader issue of animal cruelty to aberrant individuals, bad "apples," and their specific gender. Indeed, the media focus on the "Swan Lady," which focused largely on this "quaint" story of care gone wrong (but good intentions maintained), is consistent with the ways that animal cruelty cases involving agriculture also get picked up, as individual aberrations within an otherwise healthy and rational system.

Peter Chen makes the point that media coverage of animal welfare issues tends to be "episodic" rather than "thematic." An episodic framing would treat an animal hoarding issue as an individualized event, a blemish in an otherwise healthy social body, a bad day in an otherwise healthy calendar. Such a framing leaves the structural and broader social explanations (gender and violence against animals) and context largely untouched: "the individualized focus can be engaging but not necessarily productive" (95). So too in the case of the CCL and the "Swan Lady"—these are media events that do not help to either draw attention to the plight of the animals prior to their "rescue" by the hoarder, nor to the conditions under which so many animals live and die as part of a larger social system that defines itself in part by its capacity to maintain institutionalized violence "cleanly" and "rationally" and away from urban spaces, as in the case of factory farms or concentrated animal feeding organizations (CAFOs).

CCLs and CAFOs: The intensification of animal keeping

In this final section, I consider the "crazy cat ladies" appearance as a cultural trope alongside the intensification of farming practices, specifically in the form of factory farms and CAFOs. Animal hoarding, "crazy cat ladies," and CAFOs are connected in key ways but principally as examples of the intensification of animal keeping within Western modernity, and by animal keeping I mean agriculture as well as the breeding of animals designated "pets." Factory farms and CAFOs are institutions that routinely demonstrate the gulf between animal welfare standards, good intentions, and a good life for animals. The CCL who is a potential animal hoarder exemplifies the gulf between declarations of animal care and actual outcomes for animals at an individual level, and the factory farm or CAFO exemplifies it at the structural or institutional level. Thus we rarely see depictions of the latter as media stories, but a great deal of the former in media and popular culture. "Animaladies" of the hoarding variety are therefore not all equal, with some more clearly marked by gender, marking some bodies and behaviors as public exemplars of wrongdoing, while others are made invisible through normalization.

The fascination with CCLs *and* animal hoarding cases is in itself rather fascinating. It is not only the CCL products that palliate and obscure the dark side of the CCL (her animal hoarding relations) but also the documentaries and texts that want to take us "inside" their homes and minds are remarkable. Animal Planet produced three seasons (thirty episodes) of *Confessions: Animal Hoarders* (2010–2012), which exposed the lives of hoarders and the animals. They maintained a fairly consistent narrative framing, starting with concern about animal cruelty, fascination with the state of mind of the hoarder, attention to the domestic scene, and then rescue (including euthanasia) of the animals. As well as TV shows and documentaries like *Grey Gardens*, there are also texts such as *Inside Animal Hoarding: The Case of Barbara Erickson and Her 552 Dogs*, which promises to take us "inside" but begins by showing that the media got there first. As Celeste Killeen explains:

> I first met Barbara and Bob Erickson from a distance, watching the shocking television reports that describe the rescue of their surviving dogs. For several days after the dramatic police action, the evening news was filled with images of wild eyed, filthy animals, some too weak and sick to move, others seemingly trapped in a frenzy of barking and clawing … Reporters bit into the story, holding on like a dog to its bone … They ran the chaotic video clips day after day, changing only their voiceovers. In time, they reported the decision to prosecute the dog owners. But no one talked to Barbara Erickson. (2009, 6)

While sympathetic to the need to consider the perspective of Barbara Erickson, the focus on her also reinforces the ways that interest in these cases is already gendered. Her husband Bob was also charged (and pleaded guilty to charges) of animal neglect and abuse. But the state of the house, the living room, the domestic squalor, and their claims to have "loved" their "babies" (the dogs), situates this story as already about the specific failures of femininity and feminized domains: the domestic sphere, the family, and care. The access to this space is also significant. It was not just that Killeen wanted to get "inside" the mind of her subject, but so did all the reporters and, they presume, their audiences, looking into the state of the living room with a morbid and horrible fascination for such a scene of chaos and suffering. These scenes are repeated in a number of "documentaries" about hoarders, where an audience get to see "inside" the domestic spaces, implicitly enjoying the opportunity to repudiate the "madness" of those whose domestic spaces have become unruly, toxic, and pathological. To some extent, the perverse pleasure associated with repudiating such a neighbor helps to reinforce correct animal care practices and also reinforce the sense that such things should never be seen (but we must look!) in the first place.

The TV footage also works as an uncanny reminder of animal activist footage that seeks to get "inside" the factory farms to expose "hidden truths" about what animal welfare standards actually mean for animals subjected to intensive farming practices. The difference between what we see in the animal hoarder's home and what we see in the factory farm is one of degree rather than kind: the animals suffer conditions of overcrowding, confinement, in ammonia-filled air, surrounded by manure, and unable to escape the company of others or express natural behaviors. But these videos rarely make it onto mainstream television. Nor are the corporate owners and operators of factory farms subjected to the speculation of writers or media commentators about their state of mind, their peculiar biographies, or their personal traumas that may have led them to bring such a horrible site of animal suffering, like the factory farm or CAFO, into being.

Agribusinesses that rely on CAFOs are defined partly by their capacity to prevent such videos or biographies, or pathologizing and individualizing language, from becoming mainstream, with the deliberate obfuscation of public awareness about factory farming a central marketing feature. Take, for example, the packaging of animal products, described by Peter Chen as routinely depicting "pastoral scenes, with farms commonly signified by barns or old fashioned windmills … idealized or archaic portrayals, rather than the realistic representations of modern production facilities" (2017, 115). These "realistic" representations would not be unlike some of the videos on Animal Planet that expose animal hoarders in domestic urban spaces. Neither factory farm nor animal hoarder's house is structured with the animal in mind, though both may well make that claim.

Consumers of televisual and textual material on animal hoarders are given enough data to repudiate the hoarder and also sympathize with the animals. Would they also then repudiate the CAFO operator if they could also see inside? Timothy Pachirat's work on this assumption (that visibility leads to political transformation) would suggest not, especially given that "seeing" it is not the same as caring to know or acknowledging the costs for animals in the long term. Chen makes the point that consumers and voters are not presented with enough data and accurate information to make informed decisions about ethical food practices or animal welfare. He goes on to point out that the ignorance that is perpetuated by sanitized representations of intensive farming (or industrial hoarding) means that the public can become easily "shocked and angry" when confronted with such images but these come in the form of exceptional events, not structures, and that this rarely lasts long because it is not backed up by either a strong base of knowledge or consistent policy changes. This is an argument made in detail by Siobhan O'Sullivan in her book *Animals, Equality, Democracy*, which also points out that democratic principles of informed debate and discussion are thwarted by a largely secretive and invisible process of food production in Australia, as well as a strong marketing emphasis on "happy" animals on idyllic pastoral settings (2015). Carnists are not told what they are buying into and as such the cultural practices or beliefs about animal welfare suggest not an expression of their assent, or a product of democratic decision-making, but a form of consumption that is conducted largely through concealment, obfuscation, and propaganda.

What is significant about the scale of this silence around the actual conditions under which livestock in Australia live and die is that it stands in contrast to the exposure of the animal hoarder, whose mistreatment of animals is broadcast across borders and depicted in newspapers, in close and personal detail in ways that are so intense and intrusive that they might be described as verging on the violation of individual privacy and ethical duties toward the vulnerable mentally ill. Given the contrast between these two kinds of seeing— not seeing into factory farms and seeing too much into the homes of the animal hoarder—it is likely that the public has had more access to the insides of an animal hoarder's house than they have had access to the factory farm from which she or he buys her eggs, pigs, chicken, and dairy products. And from this exposure to the animal hoarder rather than CAFO operator, consumers may well be more inclined to worry about the neighbor who has too many animals in their house or yard than those high-rise sheds on the edge of town where thousands of animals are kept in windowless sheds. Given the media scrutiny of one form of intensive animal keeping over another, it is not too far-fetched to speculate that this might also skew our attention (and the attention of animal welfare officers) as to what counts as events to be alarmed about and those that may provoke disgust but not accusations of dysfunction. The difference in the quality of interest and the display of the "madness" of the animal hoarder (woman, CCL), but not factory farmer (faceless CAFO), attest to the deeply held social norms around animal agriculture that determine that it is acceptable to confine, restrict, and forcibly breed large numbers of animals that are designated livestock, but not acceptable to confine or restrict large numbers of animals designated "pets."

When the lines between domestic pets and agricultural livestock become blurred, as in the case of puppy farming, the norms by which we measure what counts as acceptable are exposed as largely arbitrary cultural formations. Public outrage that comes with puppy farming scandals attests to the idea that it is simply not acceptable to treat dogs in a way that pigs routinely are. Along with the pigs, we may well object to this arbitrary structural distinction. The arbitrariness does not translate into "easy to change," rather it means that they are prone to the sort of constant "reiteration," reinforcement, and mobility that stereotypes require. The depiction of animal hoarding cases and puppy farm scandals *as scandals* assists in putting the arbitrariness to one side momentarily, because they draw attention back to the cultural line drawn between domestic and livestock animals. Putting it another way, it's as if outrage on behalf of the puppies comes at the expense of the pigs. The CCL might be said to work in a similar way: interest in her might come at the expense of our interest in factory farmers. Distracted by her behaviors, we may fail to notice how she has come to stand in for, stand in front of, the normalization of intensive animal keeping—which (whether it be described as hoarding or not) is rarely predicated on what is the best for animals.

It is also possible that we are seeing more mainstream depictions of the CCL, this "absurd anti-heroine," at the same time that factory farming, though still largely invisible, is responding to an "increased concern for the welfare of animals" (Chen 2017, 84). This is partly the result of animal protection organizations such as Animals Australia, an organization whose support base is overwhelmingly made up of mature women,[1] much like the demographic described for animal hoarders. The dominance of women in animal protection organizations (something observed also by Munro 2001) alerts us also to the ways in which the "crazy cat lady" may well serve to trivialize and dismiss the work and political commitments of animal advocates in advance, by stigmatizing their interests in animals as potentially pathological.

Despite the benefits of the animal turn for feminist politics, and the apparent acceptance of women's independence as signaled by cat-love rather than human marriage, the persistence of the CCL as a stereotype with which to deride women's care of animals remains, especially those who are animal advocates pushed to the margins of what the mainstream counts as rational political behavior. The "crazy cat lady" has become such a key figure or symptom in exposing the gulf between care and cruelty precisely because these elements persist in much wider and institutionalized forms through intensive factory farming. The CCL is a woman of many talents: she declares her allegiance to cats; she makes a joke of and also points to both animal hoarding and women's advocacy on behalf of animals; she provides an opportunity for good intentions to obscure cruelty; and she operates as a scapegoat who can be repudiated publically while intensive animal keeping such as what we see (or don't see) in CAFOs or factory farms gets off the hook. She serves a much broader purpose as a cultural trope for a variety of "animaladies."

Acknowledgments

I would like to acknowledge Anne Fawcett's expertise and help, pointing me toward the literature on animal suffering caused by animal hoarding cases. The chapter benefited greatly from her insights into veterinary scholarship on hoarding. I thank the audience and organizers (Melissa Boyde and Alison Moore) of the "Beyond the Human" Symposium held at the University of Wollongong in 2015, where I presented an earlier draft. I would also like to thank Annie Potts for the socks, she knows why.

Note

1   Peter Chen notes that a survey of individual Animals Australia members in 2000 found 74 percent women with a mean age of 51 (Chen 2017, 179).

References

Adams C. and Donovan J. (1995), *Animals and Women: Feminist Theoretical Explorations*, Durham: Duke University Press.

Arluke, A. and Killeen, C. (2009), *Inside Animal Hoarding: The Case of Barbara Erickson and Her 552 Dogs*, Indiana: Purdue University Press.

Bhabha, H. (1994), "The Other Question: Stereotype, Discrimination and the Discourse of Colonialism," *The Location of Culture*, London: Routledge.

Chen, P. (2017), *Animal Welfare in Australia*, Sydney: Sydney University Press.

"Confessions: Animal Hoarders," *Animal Planet* (USA TV) 2010–2012.

Harrison, R. (1964), *Animal Machines*, London, Vincent Stuart Publishers.

Joffe, M., O'Shannessy, D., Dhand, N. K., Westman, M., and Fawcett, A. (2014), "Characteristics of Persons Convicted for Offences Relating to Animal Hoarding in New South Wales," *Australian Veterinary Journal* 92: 369–375.

Krasner J. (2017), "Cat Food in Camelot: Animal Hoarding, Reality Media, and Grey Gardens," *Journal of Film and Video* 69: 44–53.

McHugh, S. (2012), "Bitch, Bitch, Bitch: Personal Criticism, Feminist Theory and Dog-Writing," *Hypatia* 273: 616–635.

McKeithen, W. (2017), "Queer Ecologies of Home: Heteronormativity, Speciesism, and the Strange Intimacies of Crazy Cat Ladies," *Gender, Place & Culture* 24 (1): 122–134.

McRobbie, L. R. (2017), "The Crazy History of the Cat Lady," *Boston Globe*, May 21. Available online: https://www.bostonglobe.com/ideas/2017/05/20/the-crazy-history-cat-lady/5DJaZf5QW0KPv8KTYBBGPO/story.html (accessed May November 1, 2017).

Morrow, B. L., McNatt, R., Joyce, L., et al. (2016), "Highly Pathogenic Beta-Hemolytic Streptococcal Infections in Cats from an Institutionalized Hoarding Facility and a Multi-Species Comparison," *Journal of Feline Medicine and Surgery* 18: 318–327.

Munro, L. (2001), "Caring about Blood, Flesh, and Pain: Women's Standing in the Animal Protection Movement," *Society and Animals* 9 (1): 43–61.

O'Sullivan, S. (2015), *Animals, Equality, Democracy*, Palgrave: London.

Pachirat, T. (2015), *Every Twelve Seconds*, New Haven, CT: Yale University Press.

Patronek, G. J. (1999), "Hoarding of Animals: An Under-Recognized Public Health Problem in a Difficult-to-Study Population," *Public Health Reports* 114: 81–87.

Polak, K. C., Levy, J. K., Crawford, P. C., Leutenegger, C. M., and Moriello, K. A. (2014), "Infectious Diseases in Large-Scale Cat Hoarding Investigations," *Veterinary Journal* 201: 189–195.

Svanberg, I. and Arluke, A. (2016), "The Swedish Swan Lady Reaction to an Apparent Animal Hoarding Case," *Society & Animals* 24: 63–77.

Vialles, N. (1994), *Animal to Edible*, Cambridge: Cambridge University Press.

Chapter 12

THE ROLE OF DAMNED AND DAMMED DESIRE IN ANIMAL EXPLOITATION AND LIBERATION

pattrice jones and Cheryl Wylie

A girl's story

I started in 4-H when I was eight years old. By the time the county fair arrived, I had just turned nine and was caught up in the excitement of packing the camper and show supplies for the animals. I was taking my beloved rabbits to the fair along with my first steer, Buddy. I remember standing by as my dad and older brother got my steer to the weight scales the first day. I remember watching everyone getting him ready for the show ring that Tuesday. I remember the nervousness of being in the show ring and the excitement of being chosen for a trophy. All of those first memories will always be there, but none will overwrite the memories of Friday, which was sale day.

We lined up in the order determined by our placements in the show ring. One by one, we walked in the small sale ring with the steers we had taken care of for the past seven to ten months. The auctioneer was loud; everything else is a blur. The instructions I'd been given rang in my head: Keep moving and smile. I was supposed to smile, but all the while tears were streaming down my face. The time seemed to crawl, but I know it was just minutes before the auctioneer banged the gavel and announced "Sold!" along with the buyer's name.

I walked through the final two chutes and watched as one of the men took off Buddy's halter and handed it to me. Buddy walked through the last gate and was gone, loaded into the truck with so many others. I was supposed to find my mother to find out who the buyer was and where they were sitting, so that we could thank them and ask if they needed anything to eat or drink. After that, I headed back to the barn to empty the buckets and pack up all the supplies into the show box for the trip home on Saturday.

This repeated itself year after year—with steers, sheep, and pigs—for nine years. I cried in every sale ring, from that first one at age nine until the last time just after I graduated high school. In some respects, I think the girls had it easier than the boys. No one other than my older brother ever told me not to cry. The men working the sale always came and walked through the last gate for me. They clearly understood the sorrow, but we all still came back the next year and did it all again.

Loving them to death

Desires for close relationships with other-than-human animals are both damned and dammed. Paradoxically, people who want to be close to animals may be more likely than others to pursue careers and hobbies that hurt animals. If we can understand how people who love animals become invested in recreations and occupations that hurt animals, we may become better able to craft apt interventions. To that end, we will examine this phenomenon through three case examples (cockfighting, backyard hen keeping, and youth programs such as 4-H and Future Farmers of America), attending to both desire and identity in each case, before turning to the question of interventions. We will argue that animal advocates will succeed in altering human relationships with other animals only insofar as we acknowledge and tap into desire in all of its queerness, urgency, and gendered complexity.

We write from the mountainous terrain of a farmed animal sanctuary that offers refuge to survivors of all three of these ways that animals are hurt by people who love them, and more. In addition to survivors of cockfighting, backyard hen keeping, and 4-H programs, VINE Sanctuary has welcomed refugees from petting zoos and pigeon racing as well as from ostensibly humane farms where "beloved" cows, goats, or sheep were allowed to starve or otherwise suffer. One of us (Cheryl) participated in 4-H and FFA programs as a child, and one of us (pattrice) devised a method of rehabilitating roosters used in cockfighting—a feat that has repeatedly brought the sanctuary to the attention of voluble participants in that blood sport. We both have had extensive conversations with women engaged in backyard hen keeping. Our reflections and recommendations are rooted in such encounters with the phenomenon of "loving them to death."

Damned and dammed desire

Some desires are damned, considered by the dominant culture to be bad and wrong. In some cases, that disapproval is warranted. The ability of a community to dampen truly destructive desires is useful, and this probably explains why social disapproval can create palpable unease. As social animals, we are hard-wired to respond to expressions of disapprobation and also to internalize the values of the group. All of this goes awry when the damned desire is a harmless wish and is particularly injurious when the yearning in question is in any way vital to well-being or sense of self. The harms done to individuals by cultural or religious antipathy to homosexual desire are well known. As one of us has discussed elsewhere, suppression of same-sex eros also leads to social and environmental ills (jones 2014a).

Most people consider kindness to animals a virtue. However, in cultures founded in the speciesist notion of an absolute and hierarchical divide between humans and other animals, over-familiarity with other-than-human animals transgresses an essential border and is therefore proscribed. Whether persecuted as witches or mocked as insufficiently sane, people seen as being in unseemly communion with animals face a range of negative social consequences.

"Human" is more than a species—it's a *status*. Preservation of that status, that identity, requires constant policing of the imagined boundary between humans and all other animals. For example, humans are imagined to be more rational and less emotional than other animals. Therefore, people censured for being "too close" to animals may be told that they are insufficiently rational or overly sentimental.

Those are charges frequently leveled at women more generally. Like damnation of homosexual desire, disapproval of the wish for authentic and egalitarian relationships with other-than-human animals tends to be gendered. Girls and women are allowed and even expected to feel affection for animals, but may moderate their sympathies for fear of being seen as excessively emotional and therefore inferior to allegedly more rational males. The social psychological phenomenon known as *stereotype threat* (Inzlicht and Schmader 2011), in which members of disadvantaged groups are anxiously aware of pejorative stereotypes and sometimes alter their behavior in order to avoid seeming to embody them, may motivate girls and women to dampen or underplay their sympathy for animals.

Boys and men are even less free to express affection for animals, both because tender feelings of any kind may be construed as weakness and because the social role of "man" includes dominion over animals. Human males are supposed to want to hunt, kill, eat, and display control over members of other species. All of these activities can be used to demonstrate masculinity, and distaste for any of them may be read as femininity. Men are permitted to dote on a favored dog or horse but only so long as they maintain the position of "master" over an animal conceived as a devoted servant rather than a peer. Such motivations may redirect friendly feelings for animals into more suitably masculine efforts to control animals. When a boy who loves and wants to spend all his time with horses grows up to be a jockey who whips horses or a cowboy who "breaks" horses, we would consider this to be a case of dammed desire.

4-H and FFA

For many rural youth in the United States, participation in a 4-H or FFA club represents a kind of rite of passage. Both founded in the early 1900s in conjunction with governmental rural development efforts; 4-H and FFA teach vocational and life skills with a heavy emphasis on agriculture. 4-H members pledge to devote hands, head, heart, and health to club, community, and country. Children as young as eight can join their local 4-H club. Some clubs have a specific focus like sewing while others allow members to take any available project. Projects include skills like cooking, woodworking, photography, and gardening, but 4-H is best known for the projects in which children care for animals who are then shown, and often sold, at county fairs. All projects are designed to give the child a base of knowledge that can be built on with increasingly difficult challenges in successive years. Projects are judged either at or just before the county fair, with the results being announced during the fair for all to hear. Club members earn a few dollars for each completed project.

Cheryl recalls:

Animal projects varied greatly. Some, like rabbits, focused on proper care and nutrition for the first years. Others were "market projects" from the outset. I took my first steer to the fair at nine years old. I spent hours, first on the weekends and then daily once school let out, teaching calves to walk while haltered and learning how to position them in the show ring. I learned to groom them, and I taught them not to fear water hoses, fly spray, or the sounds of cars. I spent hours and hours teaching them to trust me, and they did.

Days at the fair started early. Rabbits had to be cleaned, fed, and watered. Steers had to be walked, rinsed, fed, and back in the barn with their stalls cleaned before the fair opened. On show day, the steers had to have a full bath. Then it was into a grooming chute to be trimmed, clipped, and combed to make them look as appealing as possible for the judge. As the assigned class for my steer approached, I was sent off to get cleaned up and changed into dress clothes. I remember being anxious as I awaited my turn in the ring. Once I entered, a thousand thoughts ran through my head, all the while I hoped that the steer wouldn't panic and bolt. There was a strange amount of pressure to do well, even though the show had nothing to do with us as kids. The placements were subjective, but I remember feeling like it meant so much.

The days between the show and sale days at our fair felt like a vacation. After everyone was fed and watered, I could run amok on the fairgrounds. The days were filled with rides, junk food, and picking up free pens in the exhibit hall. So long as I checked the animals' water and checked in for dinner, I was free. That made me look forward to fair week all year.

The last days of the fair were sale days. The times were posted, sale orders announced, and one by one the barns would be emptied. Those last two days every year were always the worst. I think that every person who cares about non-human animals has cringed to see the animal transport trucks on the road. Every year for nine years, I put my friends on one of those trucks. Then, I had to go find my mom who was watching the sale to find out who the buyer was and thank them. In nine years, I don't think I ever did it without tears in my eyes and a lump in my throat.

Originally called Future Farmers of America, FFA began as a vocational program for high school students but now can be joined by children as young as twelve. As with 4-H, club members pursue a variety of projects including but not limited to animal agriculture. Coursework might include learning about the different breeds of farmed animals but might also include woodworking or basic welding. FFA also sponsors public speaking contests, parliamentary procedure teams, and competitions judging everything from soil quality to dairy products. As with 4-H, participants take projects to the county fair, where they might win awards. Projects include horticulture, woodworking, and so-called "production projects" involving sheep, pigs, steers, or veal calves.

Cheryl recalls:

When I made out my schedule for high school, joining FFA and taking the associated class, Vocational Agricultural Education, seemed like a given. For incoming freshmen, it was a popular class, not only for its reputation of being an "easy A," but also because it was worth more credit hours than any other course. In my freshman year about half of the class took some sort of animal project to the fair. Those of us still participating in 4-H clubs had to decide early in the year not only which animals we would take but also which club each animal would fall under. Often the decision was based on which group had the earliest show times or the least competition per class. Every decision that we made during that time was based upon getting "the best return on our investment"—a goal encouraged by advisors and parents alike.

During my four years in the FFA I participated in the "breeding ewe" project in which students each take a pair of pregnant ewes, caring for them for the duration of their pregnancies and monitoring them through the birth of their lambs. Often the sheep would have multiple lambs, and I often found myself stumbling out to the barn at 2am in January to bottle-feed the lambs. Our advisor would come out within a few days of the birth to dock tails and castrate the lambs by banding. The students were responsible for making sure infection didn't set in and that the sites healed properly.

After weaning, the FFA advisor came and collected the ewes as well as one lamb per sheep. We were left with the remaining lambs to raise for the fair. I literally raised these lambs from birth, cared for them for seven months, and then sold them at auction to the highest bidder. For ten years, I cried every time I was in a sale ring. It never got easier, and no one around me ever seemed to understand that fact. The guys never seemed to cry, probably because they would be mocked if they did. Even many of the girls never seemed to shed a tear. More than once, I was accused of "faking it" in an attempt to get sympathy and a higher price from the buyers. My classmates seemed to become more callous every year. Looking back, it seems to me that we were being taught to be little sociopaths who felt nothing while betraying those we had taught to trust us.

Fair week was notorious for reckless behavior among my fellow FFA members. Despite ours being a "dry" fairgrounds, alcohol was never in short supply. Drinking was a recreational activity in the evenings and seemed to intensify year after year. The deputies assigned to patrol looked the other way unless one of us did something dangerous or the barns were loud after midnight. The final night, after the last of the sales had concluded and many of the barns were empty, it became a free-for-all of unsupervised high school students. For me, that raucous last night was an escape from the knowledge that I would go home the next day to an empty barn.

4-H and FFA shape rural youth. In addition to valuable skills like sewing, welding, horticulture, or woodworking, participants learn the value of community service. They develop the habit of seeing projects through to completion and may also develop leadership skills such as public speaking or how to chair meetings. Awards and degrees boost confidence that helps participants navigate the difficulties of adolescence. Club participation can bring the sense of belonging that is so vital to young social animals. And yet, at the same time, children who love and want to spend time with other animals are traumatized anew each year by the experience of sending beloved friends to slaughter. Over time, many of them develop emotional callouses that then allow them to become the next generation of beef farmers or pork producers.

This dispassion may foster a more widespread malaise. Cheryl recalls a dampening or blunting of all emotions, not just the sadness, as well as an unspoken ban on talking about any of the doubt, guilt, or other feelings children might feel upon sending their friends to be killed. When she says "we didn't talk about it," she means not only that her own family didn't talk about it but also that children didn't talk about it among themselves and none of the adults involved helped them to process the feelings flooding their bodies. They were dammed.

Cheryl's own recollections of these childhood experiences have many of the hallmarks of traumatic memory, with all of its nightmarish fragmentation and distortion, as if she were always walking around that show ring, trying unsuccessfully not to cry, as the auctioneer's voice blared. When she tries to open the door to more memories, it's as if something is physically blocking her from doing so. This is not surprising. Being forced to participate in killing loved ones is a kind of torture that leaves people feeling damned. Cheryl has found a way, by devoting her life to caring for animals, to assuage the lingering guilt and shame. But many survivors of this culturally approved form of child abuse repeat the pattern by subjecting their own children to it. The accumulated callousness and defensiveness may help to explain not only the perpetuation of animal agriculture but also many of the other cruelties for which the US "heartland" is internationally infamous.

Cockfighting

In a typical cockfight, men thrust two aroused roosters into unnaturally close proximity. Prevented from flight, they fight. Surrounded by shouting apes and flooded by adrenaline, each rooster concludes correctly that the only way out of this emergency is to attack. In between rounds, their handlers care for them tenderly, sometimes sucking bloody mucus from their beaks with their own mouths, even though it was they who put the birds into this mortal danger. The fight concludes only when one bird dies, at which point his handler mourns while the handler of the battered victor crows as if he himself has prevailed in battle and the gamblers begin to place their bets on the next fight.

As a pastime dating back to antiquity and spanning the globe, cockfighting has proved to be both persistent and culturally meaningful in a wide variety of settings. Here in the United States, cockfighting is illegal in all fifty states and a felony in forty of them. Unlike many other animal cruelty laws, the ban on cockfighting is actively enforced, mostly because fights are also sites of illegal gambling and associated violence. Every year, hundreds and sometimes thousands of birds are seized by authorities enforcing anti-cockfighting statutes, their numbers evidence of the continued vitality of whatever impulse leads men to abuse birds in this way.

While non-participating attendees of cockfights may be drawn by the drama of violence or the thrill of gambling, men who spend much of their own time and money raising roosters and transporting them to fights despite the risk of fines, jail time, and the social penalties of a felony conviction must be vested in the endeavor for deeper reasons. Chief among these is masculinity, but we believe that the wish for relationships also animates men who might demonstrate their virility by any number of legal and less costly ways.

In his essay entitled "Gallus as Phallus," Alan Dundes cites and summarizes numerous studies that demonstrate the association between cockfighting and masculinity within particular cultural settings, arguing that "it is always a mistake to study data from one particular culture as if it were peculiar to that culture if comparable, if not cognate, data exist in other cultures" (2007, 293). We agree that the cumulative scholarship on cockfights does suggest that they are what one of us has called "spectacles of stylized masculinity" (Jones 2010, 191) wherever they appear and even when participants insist that the activity is primarily an expression of national or ethnic identity. (Since maleness tends to be an implicit element of full citizenship, a focus on masculinity does not contradict such claims.)

Our sanctuary was the first to devise a protocol for rehabilitating roosters formerly used in cockfighting, so that they can be integrated into flocks rather than euthanized. Periodically, this fact finds its way into a television or newspaper story, at which point we begin to receive calls and messages from men who proudly participate in cockfighting and want us to know how wrong we are. The ensuing dialog tend to follow a pattern: The man begins in (1) a patronizing tone, in which he (2) presumes to know more than we, often on the basis of far less experience with roosters, and then (3) informs us that it is impossible to do what we have, in fact, done. If we contradict this in any way, then (4) the tone becomes surly and overtly sexist. There may be unappeasable demands for proof or charges of fakery, but inevitably (5) we are accused of abusing the roosters by robbing them of the opportunity to express their naturally combative personalities.

In the context of the current inquiry, the degree to which these men who abuse roosters describe themselves as *protectors* of roosters whose right to self-expression might be violated by emasculators like us is particularly telling. On the surface, this seems like a simple displacement in line with the use of roosters as avatars of masculine identity: they feel threatened by feminists, and so it feels to them like roosters are threatened by the ecofeminist animal sanctuary. At the same time, these men do seem to sincerely see themselves as caretakers rather than abusers of roosters. They will go on at great length about the specialized diets and other amenities provided to birds in their possession. They will insist that it breaks their hearts to see a rooster die in combat but that they know this was what the bird himself wanted. They seem sincerely not to understand, and not to be able to take in if the facts are explained, that their own interventions—which always include raising roosters apart from flocks in which they could learn the social skills necessary to resolve conflicts without bloodshed and which also sometimes include injecting birds with testosterone or amphetamines—produce the aggressive behavior they see in the cockfighting ring. They feel certain that they are not hurting birds and that, indeed, nobody loves roosters more than they do.

All of this is entirely consistent with the reports of anthropologists and other researchers who have spent time in communities of cockfighting enthusiasts (e.g., Geertz 1972, Maunula 2007). Men who raise roosters for fighting often spend vast amounts of time with the birds, expend significant sums on their care, and really do grieve when they die in the ring. Many develop close relationships with favored roosters, who they sometimes treat as companion animals. Thus, we can speculate that in communities where cockfighting is prevalent, this hobby might offer an appropriately "masculine" outlet for boys and men who like birds, or animals more broadly, and want to have relationships with them. We also notice how much care-taking goes into cockfighting and the degree to which cockfighters seem to enjoy tasks that would, in other contexts, be classified as feminine.

Backyard hen keeping

In the United States in the past decade, the project of keeping backyard hens for eggs has grown from a quirky hobby to a full-fledged fad complete with specialty publications such as *Backyard Poultry Magazine*. While this may seem to be a fairly benign form of animal exploitation, particularly in comparison to confinement of hens in egg factories, the hens themselves often suffer and there are several negative knock-on effects.

Backyard hen keepers usually start with chicks purchased from hatcheries. Mechanically hatched chicks come into the world looking for mothers who will never appear. The incessant peeps that seem so adorable to people are actually distress calls of youngsters crying to be sheltered under parental wings. This is only the beginning of their anguish. Whether shipped in bulk to farm stores or mailed in boxes to individuals, chicks endure terror and disorientation, potentially injurious jostling, and life-threatening dehydration en route to consumers who may have no idea how to care for healthy chicks much less be equipped to care for those disabled in transit. After landing in their new location, the chicks face an uncertain fate. Will their captors provide appropriate food and enough fresh water? Will their coop provide adequate shelter in all weather? Will they have sufficient outdoor space to maintain healthy bodies and social relations? Will they be handled roughly or kindly? What will happen if they don't lay enough eggs? What if they don't lay any eggs at all? What if they turn out to be roosters?

For every chick sexed female and sent off to become an egg producer, a hatchery kills one chick sexed as male. Hatcheries dispatch "worthless" males as quickly and cheaply as possible. After a sorting process that is itself traumatic, male chicks may be gassed, dropped into a wood chipper, or simply tossed on top of one another in a trash receptacle and left to smother or die of dehydration.

Sexing chicks is an inexact process, especially when endeavored at speed. Inevitably, some of the chicks shipped out as hens turn out to be roosters. This may not be clear to their naive keepers until they begin crowing, at which point a literally life-threatening emergency may arise. Some backyard hen enthusiasts will not tolerate roosters. In other cases, neighbors complain. Many residential areas ban roosters. Some hen keepers execute unwanted roosters, either directly or by passing them along to a farmer or family member willing to kill. Others seek to place them at a shelter or sanctuary.

As a sanctuary nationally known for our work with roosters, we have received hundreds of calls, email messages, and social media inquiries—at least one each week for more than ten years—from women seeking to divest themselves of roosters inadvertently obtained in the course of keeping hens for eggs. Here is what we have learned about and from these women:

1. They *are* women.

Apart from the occasional husband calling on behalf of his wife, virtually all of the backyard hen keepers who contact us are women. This is consistent with the long history of the exploitation of chickens for their eggs—perhaps the only process of "domestication" primarily implemented by women—as well as with traditional rural gender roles assigning the labor of hen keeping and egg collection to women and children.

2. Their motivations tend to be other than economic.

Factor in the costs of a coop, fencing, hatchlings, bedding, and feed, and it's easy to see that this adds up to more than an average household would spend on eggs. While we have sometimes heard from women who keep enough hens to make some money by selling eggs, most backyard hen keepers are paying much more for the pleasure of doing so than they save by not buying eggs.

3. Both desire and identity seem to be factors in their attraction to hen keeping.

Most backyard hen keepers seem to genuinely enjoy and value their relationships with chickens, so much so that it seems safe to infer that some wish for such relationships motivated the choice to begin keeping hens. Some subset of backyard hen keepers state that they began to keep hens in order to be able to consume eggs without being complicit in cruelties such as battery cages. This suggests both care for chickens and the wish to be a "humane" person. A smaller subset express satisfaction at producing their own food or engaging in animal husbandry, even sometimes going so far as to refer to their homes as "homesteads." Members of this subset also sometimes boast of being humane in comparison to the egg industry, but here the greater emotional investment seems to be in the identity of self-sufficient homesteader.

4. Their attachments to the birds in their care sometimes seem shallow or conflicted.

Most women engaged in backyard hen keeping express affection for the hens they call "girls" and do seem to feel attached even to the roosters they are trying to place at the sanctuary. At the same time, they will sometimes joke about "the soup pot" or otherwise make light of the fate awaiting most unwanted roosters. Expressions of affection tend to be sentimental or patronizing rather than indicative of empathic fellow feeling, as though the birds are indulged children rather than respected friends. In both the callous jokes and the treacly expressions of fondness, we discern a distancing consistent with maintaining the elevated identity of human.

5. They tend to display displacement.

Women hoping to place roosters with us often become defensive or even hostile when told that their hobby is the reason no shelter or sanctuary has room for more roosters. Most reject the recommendation that if they want to help roosters, they should stop buying chicks. Some tell us that we are rude for suggesting such a thing. Many rebuke us or sanctuaries in general for our failure to meet the demand for refuge that they themselves have created. We cannot count the number of times a backyard hen keeper calling or writing about a rooster has said that she doesn't want him to "end up" as dinner or be "sent to slaughter," as though she herself would not be the person responsible for that outcome. (We are not here talking about instances where there really is a husband or neighbor threatening to kill the birds.) This doesn't seem to be a conscious maneuver. Like men engaged in cockfighting, these women seem sincere in their feeling that they are protectors of birds without awareness that they are the cause of the menace.

Lessons from LGBTQ liberation

Recent years have brought what seems to be a sea change in public opinion concerning LGBTQ people. In many parts of the world, a goal that once seemed well beyond the reach of what could be hoped for—same-sex marriage—is now the law of the land. This remarkable story of rapid social change may be interpreted in many ways. Since we are interested in the motive power of eros (jones 2014a), we will focus on that aspect of the trajectory from sexual outlaws to wedding cake figurines.

Some LGBTQ people, despite sometimes literally lethal repression, openly and unashamedly proclaimed their romantic and sexual same-sex relationships. These visibly queer folk, along with the friends and family who expressed solidarity with them, collectively created an alternative source of social approval for LGBTQ people whose urgent desire for such approval (itself a kind of relationship) had dammed their longings for same-sex partnerships. Over time, this self-perpetuating process brought more and more LGBTQ people, and the friends and family members who continued to love them, into view. Importantly, this brought more and more LGBTQ people for whom social approval is an urgent wish into the fold of those openly agitating for LGBTQ rights. Marriage represents precisely the kind of social approval desired by those whose same-sex desires had been dammed by social disapproval and therefore became a focus of LGBTQ activism despite its relative unimportance to many of the nonconformists whose early insistence on expressing LGBTQ desire *despite* social disapproval set the process going.

But what about the desires of the non-LGBTQ "public" whose opinion seems to have so rapidly shifted? Some and perhaps most people always have felt affectionately toward their LGBTQ kin. Mothers have always loved their gay sons. Fathers who have felt bound by the demands of masculinity to disown such sons have felt extreme anguish while doing so. Siblings have felt torn between devotion to a parent and love for a peer. The LGBTQ movement has offered liberation for them too, both by offering that alternative source of social approval and by giving them something to *do* with their love for the LGBTQ people in their lives. This, too, may help to explain the rapidity with which marriage equality became real. Working for that goal was something specific and understandable that straight "allies" (many of them married) could do.

The wish for close relationships with other-than-human animals is like same-sex desire in that it is both damned and dammed. But people who love animals are also like the friends and family members of LGBTQ people, in that this is an affection for members of a subordinated class and may be the cause of some conflict. This may be one more reason to "queer" animal liberation.

Queering animal liberation

Since 2002, our sanctuary has been among a small but growing number of organizations "queering" animal liberation by drawing attention to intersections between speciesism and anti-LGBTQ bias, by promoting veganism and animal rights within LGBTQ communities, and by drawing inspiration from LGBTQ liberation movements. From our own cheeky efforts to encourage people to "eat the rainbow" (of fruits and vegetables) to a splashy queer kiss-in staged by Collectively Free to protest a fast-food chain known for both homophobia and animal abuse, these efforts have spanned the spectrum of activist tactics. To be true comrades in a shared struggle, we seek to enlist queer eros in the cause of animal liberation.

To do that, we need to do more than explain intersections or promote veganism in LGBTQ communities, although those are both essential things to do. We need to do more than swipe tactics like kiss-ins from LGBTQ activist history, although that is also something that it will be sometimes very useful and appropriate to do. To truly tap into the spirit of LGBTQ activism for the purpose of animal liberation, we will need to tune into desire and attune our interventions so that they work with rather than against the heartfelt wishes and felt identities of the people whose behavior we hope to change, most of whom were once children who loved animals and many of whom continue to claim to love animals today.

Backyard hen keepers bristle at the very idea that their hobby hurts birds because they see themselves as kind people. They reject the suggestion that they stop buying chicks because they want to be in relationship with birds. We are more likely to succeed if we work with rather than against their desires and self-identifications. Instead of telling women currently engaged in hen-keeping to *stop* buying chicks, what if we encouraged them to *start* rescuing older hens? Backyard bird keepers might even be willing to stop eating eggs altogether if offered the attractive alternative of running a "micro-sanctuary" and consequently being seen as an especially humane person. Since people tend to become vested in their projects, we may have even more success by making nonexploitive activities more attractive to would-be backyard hen keepers *before* they buy their first batch of chicks. From setting up a backyard refuge for wild birds to becoming a wild bird rehabilitator, there are so many things a woman who wants to be close to birds could do! Animal rights groups could join with organizations devoted to wild birds to advertise such activities. Similarly, organizations that already publicize the deliciousness and cost-effectiveness of veganism might also market the sensory and economic pleasures of planting, tending, and harvesting fruit and vegetable crops. Women drawn to the "homesteader" identity might respond well to a challenge to get the most food out of every square foot via veganic gardening.

As a semi-underground activity practiced by boys and men who often see it as a cultural legacy, cockfighting is a special case, but the principle of offering other ways to satisfy damned desires still applies. Here, alternatives will need to be region-specific and offered by trusted entities within the concerned communities. We cannot guess what the most attractive alternatives might be, but we can trust that there are people in those communities who hate cockfighting and would be happy to help boys who love birds discover more wholesome activities. Finding and funding those people, rather than devoting more resources to the failed law enforcement approach, ought to be a strategy of national organizations devoted to ending cockfighting. Within regions and communities where cockfighting persists, local animal protection agencies could make common cause with local organizations involved in any way with the project of defining masculinity less violently. In the absence of such resources, local animal shelters and the like could make a special effort to create volunteer opportunities and youth projects that offer boys and men the opportunity to truly become protectors and caretakers of roosters and other animals.

Concerning youth programs like 4-H and FFA, the task is easier yet harder. As noted above, those programs already offer children and teens a wealth of opportunities to learn while growing. However, because those programs were created with the specific purpose of creating the next generation of farmers, their animal-oriented projects tend to teach callousness to children who want to be friends with animals. Since rural, regional, and even racial identity can be bound up with the continuation of particular farming practices (Jones 2014b), parents may be particularly vested in such projects. Nonetheless, animal shelters and sanctuaries in rural regions could partner with local 4-H and FFA clubs to offer alternatives. For example, instead of raising bunnies, children and their families could foster litters of rescued puppies or kittens. Older participants could learn how to train dogs to help them become adoptable, with a successful adoption rather than a sale being the pinnacle of the project.

For children and teens who do intend to become farmers and are motivated by that hoped-for identity, "production" projects need not involve animals and can be designed to stimulate creativity in service of sustainability. 4-H and FFA already offer some projects involving crops. These could be expanded to focus on unusual or uncommonly high-yielding crops, with an emphasis on challenging future farmers to drawn upon the exuberance of plant varieties to feed the world in a changing climate. Similarly, as demand for plant-based meat, milks, and cheeses grows, 4-H and FFA participants can be challenged to exercise their creativity in that direction. Or, they could be challenged to emulate George Washington Carver (who devised so many uses for peanuts) by coming up with new uses for traditional crops like peas.

Regardless of whether they partner with 4-H or FFA, animal sanctuaries in rural regions have an important role to play in offering alternative not only to those programs but also to rural recreations such as hunting and fishing. The longing to get out into the woods, the pleasure of searching and finding, the flush of victory after succeeding in an arduous outdoor endeavor—all of those wishes and more can be met by on-site activities such as rock hunting, bird watching, insect identification, and forestry. Volunteer programs for youth can offer children opportunities to care for and get to know farmed animals who can remain their friends, because they will never be trucked away.

Cautions and questions

Every year in New York City, proud vegans stage a Veggie Pride Parade. Organizers of the event probably imagine that they are justly and unproblematically joining the numerous celebrations of ethnic identity or LGBTQ pride. In some ways they are right. Because of the cultural association of masculinity and meat-eating, male vegetarians and vegans really are frequently called pejorative names suggesting that they are gay. The wish for close and equitable relationships with other-than-human animals really is a damned desire in a way that really is related to homophobia. And it certainly is true that many people consider "vegan" to be a central element of their identities.

At the same time, there's no dearth of pride among vegans, so we probably don't need a parade. As we queer animal liberation, we will need to be careful not to adopt the trappings of LGBTQ activism without tapping into the motive power of desire. Parading vegetables *is* a charming spectacle. Parading fruits (pun intended) might be even more so. How might we use such spectacles to highlight the *fun* of being friends with animals rather than evoking the stereotype of the prideful vegan? And what about the prideful vegans? Like it or not, "vegan" as identity rather than practice is here to stay. In theory, visibly vegan people can collectively create an alternative source of social approval for the damned desire to be in better relationships with animals. In practice, the off-putting behavior of some of the most visibly vegan people sometimes has had the opposite effect. While there may be some short-term utility in efforts to rescue the word, "vegan" identity cannot help but center the human. We would like to see animal advocacy recenter itself on relationships.

Solidarity for every buddy

Lori Gruen (2015) rightly stresses the potential role of empathy in animal advocacy. While gendered notions about feminine care versus masculine efficiency can create an illusion of opposition, we believe that empathy can lead to greater efficacy, particularly in cases where people who say that they love animals participate in activities that harm animals. We struggle to maintain empathy with people in such situations, yet we cannot help to rehabilitate their own empathy for animals unless we understand the processes by which they learned to be callous upon entering into a project motivated by sympathy for animals. Luckily, empathy is a hallmark of healthy relationships, and so efforts to promote empathy will necessarily promote healthy relationships between human and other animals and vice versa.

Imagination is an essential element of empathy. Here too, we are in luck. People who love animals may pursue socially acceptable recreations and occupations that hurt animals because they cannot imagine other ways of being close to animals. Therefore, anything we can do to stimulate imaginative capacities will tend to both increase empathy and lessen rote participation in injurious activities. Not coincidentally, virtually every social problem we currently confront also calls out for more empathy and more imagination, so any success we might have in fostering those capabilities will help people as well as other-than-human animals.

Everybody wants to be happy, and so does every Buddy. Mutually affectionate and respectful relationships with other-than-human animals happily blur the all-important boundary that segregates and elevates "humanity." That divisive social construct—"human"—also fosters intraspecies inequalities such as ableism, racism, and sexism (Wynter 2003; Ko and Ko 2017). Therefore, sabotaging speciesism by fostering closer and more egalitarian cross-species relationships may have a more broadly salutary effect by undermining ideologies of social injustice.

Another girl's story

VINE's Pasture Pals program offers rural youth opportunities to visit and volunteer at the sanctuary. In each session, children and their caregivers receive a humane education lesson on a topic such as respect for differences followed by time spent visiting with sanctuary residents while helping with age-appropriate chores. In 2017, Willie the turkey and Domino the alpaca appointed themselves the official Teacher's Aides of the program, greeting and guiding program participants. Sadly, Willie died of a heart attack shortly before Thanksgiving, a US holiday sometimes jokingly called "Turkey Day" because of the tradition of consuming turkeys on that day. One Pasture Pals participant was so distressed to hear of his death that she prepared a poster presentation on turkeys, including a section entitled "Why I Loved Willie," as a school project. And so, instead of the usual jokes about eating turkeys, her classmates heard about the amazing capabilities of turkeys, saw pictures of several named individuals at the sanctuary, and—most importantly—

learned that their own valued friend considered a turkey to be her valued friend. They mourned with her, and no one told her not to cry.

Acknowledgments

This chapter is dedicated to the memory of young Cheryl's first steer, Buddy, and of every buddy any child has been forced to kill or surrender for sale.

References

Dundes, A. (2007), "Gallus as Phallus: A Psychoanalytic Cross-Cultural Consideration of the Cockfight as Fowl Play," in S. J. Bronner (ed.), *The Analytical Essays of Alan Dundes*, 285–316, Louisville, CO: University Press of Colorado.
Geertz, C. (1972), "Deep Play: Notes on the Balinese Cockfight," *Daedalus* 101 (1): 1–37.
Gruen, L. (2015), *Entangled Empathy: An Alternative Ethic for Our Relationships with Animals*. New York: Lantern Books.
Inzlicht, M. and Schmader T. (eds.) (2011), *Stereotype Threat: Theory, Process, and Application*, Oxford, GBR: Oxford University Press.
Jones, p. (2010), "Harbingers of (Silent) Spring: Avian Archetypes in Myth and Reality," *Spring: A Journal of Archetype and Culture* 83: 185–212.
Jones, p. (2014a), "Eros and the Mechanisms of Eco-Defense," in Carol J. Adams and Lori Gruen (eds.), *Ecofeminism: Feminist Intersections with Other Animals and the Earth*, 91–106, New York: Bloomsbury.
Jones, p. (2014b), *The Oxen at the Intersection: A Collision*, New York: Lantern Books.
Ko, A. and Ko, S. (2017), *Aphro-Ism: Essays on Pop Culture, Feminism, and Black Veganism from Two Sisters*, New York: Lantern Books.
Maunula, M. (2007), "Of Chickens and Men: Cockfighting and Equality in the South," *Southern Cultures* 13 (4): 76–85.
Wynter, S. (2003), "Unsettling the Coloniality of Being/Power/Truth/Freedom: Towards the Human, After Man, Its Overrepresentation—An Argument," *The New Centennial Review* 3 (3): 257–337.

Chapter 13

DUCK LAKE PROJECT: ART MEETS ACTIVISM IN AN ANTI-HIDE, ANTI-BLOKE, ANTIDOTE TO DUCK SHOOTING

Yvette Watt

An activist is someone who cannot help but fight for something. That person is not usually motivated by a need for power or money or fame, but in fact is driven slightly mad by some injustice, some cruelty, some unfairness, so much so that he or she is compelled by some internal moral engine to act to make it better.

—Eve Ensler (2013, ix)

Introduction

On March 5, 2016, just before sunrise, an event unfolded in a remote location on the east coast of Tasmania, Australia. In the planning for nine months, it was something of a mix between Werner Herzog's *Fitzcaraldo* in production and Stephan Elliott's *Priscilla Queen of the Desert* in form. The scene involved a troupe of dancers in hot-pink, sparkly tutus and pink leggings performing a specially choreographed routine to music from *Swan Lake*, on a stage floating on a wide lagoon (see Figures 13.2 and 13.3). Meanwhile, a team of hot-pink camo-clad people with pink sparkly flags made their way across the lagoon, some on foot, others in kayaks towing hot-pink ducks behind them. It was an incredibly still, warm morning, and the lagoon was like a mirror. There was a buzz of excitement that the long-awaited morning had arrived, and a delight that there were media there to record it. But there was also a strong undercurrent of anxiety and anger tied to the very reason that we were all there: the opening of duck shooting season in one of the last places in Australia where recreational duck shooting is still legal.

Since 2003 I have been going to Moulting Lagoon on the east coast of Tasmania for the opening of duck shooting season, as a member of the duck rescue team. According to Tasmania Parks and Wildlife Service, Moulting Lagoon is home to "the largest concentration of black swans in Tasmania, with an average of between 8,000 and 10,000 swans living in the lagoon" (2016, 1). The lagoon gets its name from the swans' flight feathers that accumulate along the shoreline during their annual molt. Despite being a Ramsar[1] listed wetland of international conservation significance, this site is open to duck shooters for three months between March and June each year, and the shooters' hides are there all year round. This site has been the focus of the campaign to end duck shooting in Tasmania for at least two decades now as it is the main public wetland accessible to shooters, with duck shooting commonly taking place on private property (for one strategy to prevent hunting see Figure 13.1).

Tasmania is one of three Australian states that still allow recreational duck shooting, with permanent bans in place in Western Australia (since 1990), New South Wales (1995), and Queensland (2005). Recreational duck shooting has never been allowed in the Australian Capital Territory. Those states that have banned recreational duck shooting have done so primarily on animal welfare grounds, due to the high wounding rate that is an inevitable result of firing shot into a flock of birds. 1,178 registered duck shooters in Tasmania are estimated to have killed 58,298 ducks in the 2017 season (Tasmanian Government Department of Primary Industries, Parks, Water and Environment Wildlife Management Branch, 16), and the RSPCA Australia estimates that wounding rates quite possibly equal kill rates (n.d.). The shooting of protected and endangered species is also a concern. There are five listed species of ducks that can be legally shot during the Tasmanian recreational duck shooting season, with all other species protected. However, even at a public site such as Moulting Lagoon, protected species, including black swans, have been found shot.

Figure 13.1 In 2008 Catherine Silcock (far left) devised these innovative windsocks, which proved to be highly effective in deterring ducks away from the shooters' guns. One of the hunters' hides is on the left-hand side. Photo: Yvette Watt.

This vast estuarine lagoon is for the most part very deep, but it has three shallow banks where the shooters build their hides. Our strategy as duck rescuers is to spread ourselves around the lagoon, either on foot or in kayaks, paying particular attention to occupied hides. We deter the ducks from the shooters' guns by waving brightly colored flags, as the ducks are highly sensitive to unusual movement and also see a broader range of colors and with more intensity than the human eye—hence the need for hides (Anderson et al. n.d.). Our strategy is simple but effective as the ducks tend to fly high to avoid the unusual color and movement, and, move, we hope, out of the forty to fifty meter range of the shotguns. We also try to rescue injured ducks that have been abandoned by shooters,[2] but the expanse of the lagoon and the fact that wounded ducks often dive underwater in an attempt to escape mean that rescuing injured birds is difficult and those that are rescued rarely survive. So our main aim is to stop the ducks from being shot in the first place. We also aim to get media coverage each year to bring the issue to the attention of the general public, in the hope that this will result in pressure on the Tasmanian government to ban recreational duck shooting in line with other states. But the remoteness of the location, the lack of suitable footage or photographs, and the increasingly tight budgets of media outlets have made getting coverage more and more difficult to achieve. The number of duck rescuers has also started to dwindle over the years. I hadn't been able to attend the opening weekend rescue for three years running and when I returned in 2015, I was shocked to find that there were only about five of us. It seemed the anti-duck-shooting campaign had lost momentum, for the activists, for the media, and for the general public.

Despite the drop in the numbers of duck rescuers in recent years, the sustained campaign at Moulting Lagoon has also seen a drop in shooter numbers. A number of hides have been burned down over the years (no one is owning up to how that happened) and a good number of these have not been replaced. The shooters, who are almost always white men, embody the kind of masculinity embodied in the Australian "bloke."[3] While there is a degree of affection attached to this term at times (you might hear someone referred to as "a good bloke"), it also refers to the type of white male masculinity with a strong rural base and a rough and at times violent manner–this is exemplified in the pejorative name of "wife-beater" that is given to the blue singlet, which is often associated with the "Aussie bloke." This type of man also has a liking for cars and guns. As Marti Kheel has observed, "The association between hunting and masculine self-identity has been a recurring theme throughout history" (Kheel 1996, 38).

Duck Lake Project takes flight

I came away from the opening weekend in 2015 with the feeling that something more needed be done to address what I had increasingly come to think of as the profound "blokey-ness" of duck shooting, the antidote for which might be something like people in tutus on a floating stage, performing to the music from *Swan Lake*. My initial thought was that the "over the top" macho culture of duck shooting would be highlighted by the disruptive and opposing presence of a similarly "over the top" feminine display of pink tutus, ballet, and classical music. I spent a few weeks considering what it would take to turn such a crazy idea into a reality. In working out how to make the project happen, I was also recognizing that this would require a new mode of working for me as an artist and would blur the lines between art and activism beyond what my work had done before. The possibilities were exciting, but also daunting.

In late March I emailed an artist and filmmaker friend, and told him of my plan, asking if he would be prepared to help me document it. His extremely enthusiastic response helped seal the deal—I felt I had now committed myself to bringing this "crazy" idea to fruition. Nonetheless, I spent a lot of the time somewhat terrified at what I had taken on, and not at all sure I could bring it off. Artist Rick Lowe discusses this in relation to his *Project Row Houses* work: "Oftentimes as an artist you're trespassing into different zones ... Oftentimes ... I know nothing. I have to force myself and find courage to trespass ... Artists can license ourselves to explore in any way imaginable. The challenge is having the courage to carry it through" (Thompson 2012, 93).

Ultimately, I just had to have faith that in some form it would happen, though it was only on the day that I had any sense of exactly how it would all come together.

From the outset I was truly amazed at the enthusiasm and commitment shown by the people I asked to be involved. Where previous artworks of mine have required only a passive engagement (a "looking") from the viewer, *Duck Lake Project* involved a large group of people actively engaging in a whole performance; the line between artist and audience was blurred from the outset; not only were there many artistic collaborators, the rescuers, the shooters, and even the police and Parks and Wildlife officers were in a sense both participants and audience. This was a truly a participatory art project—I had, in the words of artist Jeremy Deller, gone from "being an artist that makes things, to an artist that makes things happen" (Thompson 2012, 32). In order to make *Duck Lake Project* happen, a large group of volunteers was required, including a choreographer, a team of dancers (all of them amateurs), costume and prop makers, people to assist with the crowdfunding campaign and design work, people to assist with catering for the event, people to bring the portable toilets, a team to document the event itself and its development over the preceding nine months, a stage designer/builder, and a large team of duck rescuers and general helpers to be there on the opening weekend. Significantly, the project also received enthusiastic support from the Tasmanian arts community, with more than sixty artists volunteering their time and talents to decorate a "decoy" as part of the crowdfunding campaign. I was overwhelmed at the unequivocal support in this regard from the vast majority of artists who I approached, with artists from around Australia emailing me to ask to be involved, and a number of artists expressing disappointment that I hadn't asked them! It was thanks in no small part to these artists that the crowdfunding campaign was so successful, raising a total of $10,000 for the project. As the day of the event drew near, however, one of the key aspects of the performance—that it would be on a floating stage on the lagoon—was suddenly placed in doubt.

Some five or six weeks prior to the event I was contacted by officials from the Tasmania Parks and Wildlife Service who had heard of our plan. In a discussion over the phone they expressed various concerns such as the possibility of causing damage to the foreshore where the stage was to be located. I explained that the stage would be on the water, not on the foreshore. They indicated that I would need to get approvals from Marine Safety Tasmania if the stage was to be floating. I checked and needed no such approval. As requested, I sent them a detailed outline of the plan for their assessment, addressing all the concerns they had raised. Eventually, just four days prior to the event, when everything was ready to go (and after two emails from me asking for a response to the outline I had sent), I was advised that we would not be granted permission to place the stage on the water and would only be issued with a permit to place the stage in a day-use car park near the lagoon. To say we were disappointed would be an understatement.

On Friday March 4, we arrived at the lagoon, set up camp, and set up the stage in the car park as required—there seemed to be no other option without risking the whole event being shut down by Parks and Wildlife officers and police. Last rehearsals took place, before we all retired to camp for an early dinner and an early night, ready to be up before dawn (shooters are allowed to start shooting an hour before sunrise and an hour after sunset). I woke early and got the coffee on to warm people up before they headed to the lagoon, and it was only then I learned that overnight, some of the activists had taken it upon themselves to move the stage onto the water. My delight at this news turned to a concern that we would be shut down before the media had a chance to film the performance. The morning was warm and still as we walked down to the lagoon in the dark. The stage had a portable floodlight at its edge, and the hot-pink of the dancers' costumes were reflected in the mirror-like surface of the lagoon. It was a stunning sight and I was extremely relieved when the media arrived before Parks and Wildlife officers did. We began the performance in the dark, the sound of the music from *Swan Lake* filling the air, while the dancers began their performance.

The dancers—Judy Blackwood, Kim O'Sullivan, Debbie Lustig, Kate Mascall, Madeleine Southey, Adam Christ, and Sophie Bullock—were seven amazing women ranging in age from their mid-twenties to mid-fifties. A mix of activists and artists, all were both nervous and excited. Debbie, a veteran duck rescuer from Melbourne, was so excited at what *Duck Lake Project* was trying to achieve that she had been flying in and out of Melbourne for the rehearsals. The special routine they had been working on for weeks with choreographer Glenn

Figure 13.2 Dancers in Moulting lagoon. Photo: Michelle Powell.

Murray, a past member of the Australian Ballet and Sydney Dance Company, was to finally be performed on location, in front of the media, and with the risk that Parks and Wildlife officers would order them off the stage when they arrived. While Kim, Debbie, and Adam had done some amateur dance training in the past, all the others were complete novices—in fact when Judy volunteered to help with *Duck Lake Project* she anticipated doing something like making sandwiches! While Debbie was terribly anxious, poor Sophie's nerves got the better of her and she couldn't bring herself to perform until much later in the morning. The rest of the troupe walked through the shallow water and climbed the stage, feeling it rock as they did, with us all wondering how stable it would be. As they took their places on the stage you could see the mood of determination take over. I pressed play and as the first strains of music drifted across the lagoon, these seven wonderful women began the routine.

The media were delighted and filmed over some time as the dancers repeated the six-minute routine. Meanwhile the large team of rescuers began to make their way onto the lagoon, as shooters made their way out to their hides. A short time later Parks and Wildlife officers arrived, but rather than stopping us, they seemed equally delighted at the scene. It all came together almost exactly as I had imagined, and I still feel emotional when I recall this scene and think of the incredible team of people who made it happen.

Art and activism

Duck Lake Project provides a useful platform to reflect on the relationship between art and activism; gender, violence, and resistance; and the nature of academic life in the context of animal advocacy. It has been close to a decade now since I was actively engaged with an animal advocacy group, but my commitment to the animal rights–based principles I first embraced thirty years ago has not wavered. My artwork has remained engaged with the ethical and political dimensions that are such a big part of the interactions between human and nonhuman animals. I have significantly less time for my art practice now, due to the demands of an academic position, but this has not dimmed my determination to use my work to promote a consideration of the treatment of animals by humans. If anything, I have had to actively blur the "line" between art and activism over the years. *Duck Lake Project* manifests a new art territory for me, opening up a world of possibilities that emerge when activism and art become entangled.

The sort of practice that *Duck Lake Project* signifies might be described as "relational aesthetics," "socially engaged art" (Pasternak in Thompson 2012, 8), "performative democracy," or "artivism" (Weibel 2013, 23). Artworks that operate within this model are generally collaborative, participatory, event-based and foreground social issues and political activism. I fell into this accidentally—I didn't even stop to consider that I was working within this mode until after the fact. But having acknowledged that *Duck Lake Project* fits within the definitions of socially engaged, participatory model of art practice I have taken an interest in the artists and writers who deal with this way of working. What is interesting about this practice is that despite the obvious potential for socially engaged art to provide a productive platform for drawing attention to issues concerning the effects of human exceptionalism of nonhuman animals, key texts and websites on this mode of practice contain few or no such examples. For example, Nato Thompson's 2012 book *Living as Form: Socially Engaged Art from 1991–2011* contains not one project that is concerned with the fate of animals at the hands of humans, despite detailing a total of 104 projects. He even lists "seminal pedagogic social movements of the last 100 years … include[ing] AIDS activism, the women's movement, the anti-Apartheid movement, Perestroika, the civil rights movement, Paris '68, the Algerian Wars" (21). This is particularly curious given that Thompson curated the exhibition *Becoming Animal* at MassMOCA in 2005. Further evidence of the paucity of coverage for art and animal activism can be found in the 700+ page book *Global Activism: Art and Conflict in the 21st Century,* published to accompany the exhibition *Global aCtIVSsm* at the ZKM Center for Art and Media in Karlsruhe, Germany, in 2013. There is a passing mention on animal rights (58), but of over 120 featured projects/events/artists/organizations, just two feature animals. However, only one of these is actually advocating for animals: the work of artist collective NEOZOON whose powerful work interrogates the complexities of human–animal relations. The other work, by Moroccan artist Mohammed Laouli, is a video work that shows an abandoned gray horse with its front hooves tied together in a square of a poor neighborhood of Rabat. Laouli has adorned the horse with wings, but rather than intending to draw attention to the suffering of animals, or even the plight of this particular horse, the horse in this work "represents a form of otherness, which, from the perspective of a different social class, is also characteristic of the people living in this neighbourhood" (Weibel, 616).

One might suppose that the lack of coverage of animal activist artists and events in such books is based on a lack of suitable examples. If so, then I would encourage artists and activists to work to remedy this. However, given the broad scope of *Global Activism*, it is curious that the many performative protests of organizations such as PETA, Anonymous for Animals, and Igualidad Animal (Spain) are given no coverage despite their alignment with the forms of "performative democracy" and "artivism" included in this book and exhibition. Further, that animal-artists with well-established credentials and reputations such as Snaebjornsdottir/Wilson, who have worked in this mode for many years, fail to rate a mention in the literature suggests that animal-themed artworks simply get less attention than those focused on human social-justice issues. This is not meant as a criticism of the many wonderful socially engaged art projects on human social justice issues that are given good coverage. But the lack of attention given to animal-themed socially engaged art projects may be yet more evidence of the animal rights movement being seen as "the orphan of the left" (Kymlicka 2014, citing Canadian author Blaire French), or the orphan of the art world, as the case may be. Kymlicka elaborated on the position of animal rights advocates being the "orphan of the left" by suggesting that the left is concerned that animal advocacy displaces human-centered social justice, that it trivializes the latter and is incompatible with movements that gain power by distancing themselves from forms of "animalization" (2014). As W. J. T. Mitchell has stated, "The question of animal rights produces a combination of resistance and anxiety because to claim rights for animals entails a revolution so profound it would shake the foundations of human society" (Wolfe, x). An artist who engages with these issues faces real challenges due to this anxiety and resistance from those who feel threatened by having their foundations (strategies, priorities, and claims for justice) shaken. While posts to Tasmanian hunter Facebook groups suggest that those who felt most threatened by *Duck Lake Project* may have been the hunters, the resistance to any challenge of the broader consequences of human exceptionalism is of course widespread. However, this should not stop us from shaking things up, but we also need to consider that if we want to engage people rather than simply upset them, the best form of shaking may be firm and gentle rather than aggressive, engaging rather than confronting.

A questioning of old modes of protest is growing, with Micah White, co-creator of the Occupy Wall Street movement, making strident calls for a change to the way activists work, paying particular attention to the ineffectiveness of street protest in achieving change. Artist and "Godmother of craftivism" (craft+activism), Betsey Greer's Craftivism website reflects White's allegation, stating "people have grown tired of so-called traditional ways of activism … They are looking for a way to connect and deepen their understanding of things.

Yelling doesn't change things, but dialogue does" (n.d.). White advocates for a form of activism that uses the electoral system via the formation of political parties as a key strategy. My views align with Greer's in believing that that the visual arts can play an instrumental role in crafting change. As Edelman notes:

> Although art is no more a bastion of democracy than elections and lobbying are, it does strengthen democracy in some respects. Because it excites minds and feelings as everyday experiences ordinarily do not, it is a provocation, an incentive to mental and emotional alertness. Its creation of new realities means that it can intrude upon passive acceptance of conventional ideas and banal responses to political clichés. For that reason art can help foster a reflective public that is less inclined to think and act in a herd spirit or according to the cues and dictates provided by a privileged oligarchy. (1995, 143–144)

In general, the strategies I employ in my artwork aim to encourage in the audience a consideration of the issues at hand rather than berate them with an anger-driven polemic. The tone varies, and can at times be quite somber, but until now I have avoided the use of graphic images of animal suffering, believing it is important to engage rather than repel the viewer. *Duck Lake Project* seeks to engage a broad range of people, using a form of "soft power" to influence opinion, engaging humor as a key strategy to bring home a serious point. But in deflecting the shock of animal suffering by using humor am I softening the blow to the point that it is barely hitting home? I would argue not. In an article posted on his *Striking at the Roots* blog titled "How Do Graphic Images Affect Animal Advocacy," Mark Hawthorne looked at two studies on the differing effects on viewers of imagery of animals that varied from graphic to fairly mild. One was conducted by the not-for-profit group Farm Animal Rights Movement using still photographs and the other by the Humane Research Council using videos. Interestingly, where the former study found that the least graphic images had the most affect, the latter showed the opposite. Through speaking with various other activists Hawthorne suggests that both strategies can be effective depending on the person, the timing, and the context, a matter supported by reader comments.

At the heart of the work of the activist artist is a hope that in some way their artwork may assist in enacting social change. While it may be a large claim to suggest that art can be the instigator of such change in and of itself, I do believe it has an important role to play in engaging people's imaginations and provoking consideration of the issues at hand. Thompson speaks of this when discussing the powerful work *Palas Por Pistolas* (Spades for Pistols), which saw the artist Pedro Reyes melt down 1,527 guns and turn them into spades, which were used to dig the soil to plant 1,527 trees. He says: "Pedro Reyes did remove 1,527 guns from the streets of Culiacan. But given the actual extent of the gun violence there, his gesture seems more symbolic than practical. And yet, symbolic gestures can be powerful and effective methods for change" (2012, 18).

The issues around communicating a message and enacting change provide significant challenges for the activist artist. In blurring the boundaries between art and activism there is the quandary of how to negotiate what Thompson refers to as the "Spectrum of Legibility" (2015, 34). As he points out:

> For many involved in the arts, an artwork must remain opaque enough to invite a proper amount of speculation and guesswork ... In activism, though, clarity is celebrated, and a cogent message can reach a wide audience and can serve as a weapon. The two ends of this dynamic, which I refer to as the ambiguous and the didactic, have long proven irreconcilable. (34)

Finding ways to counter the resistance to pro-animal work (for want of a better phrase) has informed my artwork for many years, but as an academic with a tenured university position, this comes with added complexities. Initially I had intended to undertake *Duck Lake Project* as a research project within my academic role. I soon realized, though, that in the current institutional climate of caution and risk aversion this would be fraught, if not impossible. Combine an unspecified and changing team of volunteers, with guns, water, and potential hypothermia, and a politically driven project aiming to bring about legislative change, and it is clear that I would struggle to get such a project through institutional ethics approval and workplace health and safety risk assessment processes, not to mention that it may have resulted in potentially unwelcome media attention for the university.[4] Given that I was eventually charged by the Tasmania Parks and Wildlife Service with not complying with a permit, resulting in a court appearance,[5] it was a wise move to undertake this endeavor as a tutu-wearing regular citizen rather than as an academic with ethics approval. While the boundaries between academic work and activism have never been secure either (feminist, queer, indigenous scholarship routinely promotes the necessity of doing both simultaneously), the violence that is at the core of *Duck Lake Project*'s object—men with guns shooting at birds—meant that it could not be conducted as a "safe" academic exercise.

Safe in the sense of keeping participants from being in danger, and also "safe" in the sense of staying within expected norms and activities (such as writing book chapters!) for academic work. The project had to blur a number of lines in order to exist: the line between artist and audience, between art, activism, and academia, and between doing "sensible" political work of advocacy in suits and being "ridiculous" in a pink tutu. It is in these different senses that I see this as a somewhat renegade project—one that pushes at a variety of boundaries, testing their permeability while risking the consequences of breaking through.

Our intention was to pressure the state government to pursue legislative change, and use the media to turn attention back to this issue that had for a while now been receding in the public's consciousness. It aimed to get the media there: and it did. The media coverage was both thorough and resoundingly positive, airing on all three Tasmanian TV news bulletins as well as on the Australian Broadcasting Commission's Victorian TV news bulletin. Press coverage was similarly thorough, with the story and accompanying image syndicated via Murdoch Press throughout Australia. Images from the performance were also used by the media for the 2017 duck shooting season, as well in coverage of the resulting court case several months later. The project successfully reignited an interest in the campaign in terms of both getting some forty-five or more people up to the lagoon for opening weekend who for the most part experienced the opening of duck shooting season for the first time, and gaining a great deal of interest from nonparticipants via social media and the crowdfunding campaign. It also aimed to counter the hypermasculinity of duck shooting, a matter that was taken up in the report by the Australian Broadcasting Commission TV and online news bulletins, while the response from blogger Matt Hayden suggests it also struck a nerve for him in his post titled "Tasmanian Frightbats[6] Dance *Swan Lake* to Subvert Patriarchal Duck Hunting" (2016). *Duck Lake Project* was a high-camp, kitsch spectacle that aimed to challenge the gendered, dominant view of the lagoon as a place for men to kill birds and was underpinned by a general concern over the domination of nonhuman animals by humans. As Jacques Ranciere writes, "In its most general formula, critical art intends to raise consciousness of the mechanisms of domination in order to turn the spectator into a conscious agent in the transformation of the world" (Bishop 2006, 83). While it is impossible to quantify precisely who the audience is in all the forms it has taken for this widely disseminated project, it would include the participants, police and Parks and Wildlife staff, and even the hunters. Whoever the audience is, their reaction and whether I can convince them of the need to end recreational duck shooting in Tasmania is really the end game, if not the swan song.

Figure 13.3 Photo: Michelle Powell.

Conclusion

Duck Lake Project is a peculiar hybrid beast of a project—part artwork, part strategy to prevent ducks from being shot, part media stunt. But the project is also more than just the event itself, having an afterlife as a two-channel video artwork, as a publicly available documentary summary on YouTube,[7] in being spoken about by me in various academic and nonacademic presentations, and more recently being featured in the revised version of of Carol J. Adams 1994 book, *Neither Man Nor Beast: Feminism and the Defense of Animals* (2018).

The incongruous scene of dancers in hot-pink tutus, performing a routine based on bird movements to the music from *Swan Lake*, on a floating stage in a remote lagoon, came about due to my firm belief in the power of art to capture the imagination and to engage emotions, and that art that does this has the potential to challenge the status quo and assist with achieving legislative change. But, has *Duck Lake Project* resulted in any change? Well, it hasn't led to the government banning duck shooting in Tasmania, but this "mad" project has had the kind of immediate and quantifiable impact that no other project I have undertaken to date has had. But perhaps the most important result of *Duck Lake Project* is that it stopped a good number of ducks from being killed and that, in my opinion, can be claimed as a real success.

Acknowledgments

The list of acknowledgments for *Duck Lake Project* is incredibly long—too long to list here. But I must acknowledge the following people who were instrumental in the event's success: my dear friend, artist and key collaborator: Christina Scott; choreographer: Glen Murray; dancers: Judy Blackwood, Sophie Bullock, Adam Christ, Debbie Lustig, Kate Mascall, Kim O'Sullivan, Madeleine Southey; stage design and construction: Jed McNeill; documentary film team: Raef Sawford and Rummin Predictions; photographer: Michelle Powell; and the vast team of people who were involved in one way or another. The generosity of the people who helped, donated, or both was something truly special.

I would also like to acknowledge the artists who produced artworks for the crowdfunding campaign here (in no particular order): Justy Phillips, Margaret Woodward, Phoebe Adams, Leigh Hobba, Ros Meeker, Michael Nay, Lorraine Biggs, Amanda Davies, Ben Booth, Matt Warren, Tricky Walsh, Mike Singe, Mary Scott, Carolyn Wigston, Mat Ward, Penny Burnett, Christine Scott, Karen Cooper, Brigita Ozolins, Ben Kluss (Jamin), Mish Meijers, Pat Brassington, Meg Walch, Tony Richardson, David Edgar, Charles Murdock, Jan Hogan, Neil Haddon, Colin Langridge, Sally Rees, Keven

Francis, Cath Robinson, Katy Woodroffe, Lucy Bleach, David Keeling & Helen Wright, Annalise Rees, Joel Crosswell, Kate de Salis, Sue Lovegrove, Julie Gough, Ian Bonde, Patrick Hall, Di Allison, Wayne Brookes, Tom O'Hern, Lucy Hawthorne, Steven Carson, Bill Hart, Nicole Robson, Pip de Salis, Louise Josephs, Antonia Aitken, Kim O'Sullivan, Tor Maclean, Yvonne Rees-Pagh, Meg Keating, Martin Walch, Erin Amor, Lucia Usmiani, and Marion Marison.

Notes

1   The Ramsar Convention on Wetlands is of international importance especially as Waterfowl Habitat is an international treaty for the conservation and sustainable use of wetlands. It is named after the city of Ramsar in Iran, where the convention was signed in 1971.
2   We don't compete with shooters to claim a duck. Not only does this risk us being charged with theft, but the shooters have a daily bag limit, so if we allow the shooter to keep a duck that is dead or likely to die this will be one less duck he can shoot before reaching his limit.
3   A "bloke" is an Australian/British slang term for a man. In Australian usage, it tends to refer to a particularly masculine archetype.
4   Interestingly, a now-retired colleague did indeed produce an artwork based on his experience of the opening of duck shooting season at Moulting Lagoon in the early 1990s, and a local artist recalls as an undergraduate being taken by this colleague on a field trip there during the duck shooting season for his video class at around the same time. It is notable that this colleague had his tires slashed, presumably by an angry duck hunter. This was almost a decade after virtually the entire art school staff went en masse in a bus to protest at the damming of the Franklin River. Times have well and truly changed.
5   The case was dismissed in early June 2017 after Tasmania Parks and Wildlife Service failed to provide any evidence to support the charge.
6   A "frightbat" is a term coined by Australian journalist Tim Blair, who since 2014 via an opinion column in the *Daily Telegraph* titled "Crown Our Crazy Queen" asks readers to vote for "Australia's craziest left-wing frightbat" (2017).
7   An eight-minute video documentary of the project is available (*The Duck Lake Project*, 2016). Available online: https://www.youtube.com/watch?v=CDIlURup9Ok

References

ABC News (2016), "Protesters against Duck Shooting Don Pink Tutus for *Swan Lake* Performance at Moulting Lagoon Season Opener," March 5. Available online: http://www.abc.net.au/news/2016-03-05/duck-shooting-protesters-don-pink-tutus-at-moulting-lagoon-seas/7223350 (accessed September 11, 2017).
Adams, C. J. (2018), *Neither Man Nor Beast: Feminism and the Defense of Animals*, New York: Bloomsbury Revelations.

Anderson, K. A., Unghire, J. M., and Coluccy, J. (n.d.), "A Bird's-Eye View." Available online: http://www.ducks.org/conservation/waterfowl-research-science/a-birds-eye-view (accessed May 22, 2018).

"An In-depth Look at the Amazing Visual Abilities of Waterfowl" (n.d.), Available online: http://www.ducks.org/conservation/waterfowl-research-science/a-birds-eye-view (accessed September 8, 2017).

Bishop, C. (ed.) (2006), *Participation*, London: Whitechapel/MIT Press.

Blair, T. (2014), "Crown Our Crazy Queen," *The Daily Telegraph*, June 24, 2017. Available online: https://www.dailytelegraph.com.au/blogs/tim-blair/crown-our-crazy-queen-in-2017/news-story/75e28dd921ccf37129fb2b18d4b84459 (accessed February 25, 2017).

Craftivism: Craft + Activism = Craftivism (n.d.), "Craftivism." Available online: http://craftivism.com/ (accessed September 12, 2017).

Edelman, M. (1995), *From Art to Politics: How Artistic Creations Shape Political Conceptions*, Chicago: University of Chicago Press.

Ensler, E. (2013), "Foreword," in J. Williams (ed.), *My Name Is Jody Williams: A Vermont Girl's Winding Path to the Nobel Peace Prize*, i–xi, Berkeley: University of California Press.

French, B. (1988), *The Ticking Tenure Clock*, New York: Suny Press.

Hawthorne, M. (2012), "How Do Graphic Images Affect Animal Advocacy?" *Striking at the Roots*, November 1. Available online: https://strikingattheroots.wordpress.com/2012/11/01/how-do-graphic-images-affect-animal-advocacy/ (accessed September 11, 2017).

Hayden, M. (2016), "Tasmanian Frightbats Dance to Swan Lake," March 5, 2016. Available online: http://www.matthaydenblog.com/2016/03/tasmanian-frightbats-dance-swan-lake-to.html?m=1 (accessed September 11, 2017).

Kheel, M. (1996), "The Killing Game: An Ecofeminist Critique of Hunting," *Journal of the Philosophy of Sport* 23: 30–44.

Kymlicka, W. (2014), "Public Lecture: 'Towards a Multicultural Zoopolis: Animals Rights, Race and the Left,'" August 11, 2014, Faculty of the Arts, University of Tasmania.

RSPCA Australia (n.d.), "What Are the Wounding Rates Associated with Duck Hunting?" Available online: http://kb.rspca.org.au/What-are-the-wounding-rates-associated-with-duck-hunting_529.html (accessed September 8, 2017).

Tasmanian Government Department of Primary Industries, Parks, Water and Environment Wildlife Management Branch (2018), *Game Tracks* 23. Available online: http://dpipwe.tas.gov.au/Documents/Game-tracks.pdf (accessed February 25, 2018).

Tasmanian Parks and Wildlife Service (2016), "Moulting Lagoon Game Reserve" (Ramsar Site). Available online: http://www.parks.tas.gov.au/file.aspx?id=19222 (accessed June 30, 2016).

The Duck Lake Project (2016) [Film]. Rummin Productions. Available online: https://www.youtube.com/watch?v=CDIlURup9Ok&feature=youtu.be (accessed September 11, 2017).

Thompson, N. (2012), *Living as Form*, Hayward: MIT Press.

Thompson, N. (2015), *Seeing Power: Art and Activism in the 21st Century*, Brooklyn: Melville House Publishing.

Weibel, P. (ed.) (2013), *Global Activism: Art and Conflict in the 21st Century*, Karlsruhe: ZKM Center for Art and Media.

White, M. (2017), "Occupy and Black Lives Matter Failed. We Can Either Win Wars or Win Elections," *Guardian Australia*, August 29. Available online: https://www.theguardian.com/commentisfree/2017/aug/29/why-are-our-protests-failing-and-how-can-we-achieve-social-change-today (accessed September 12, 2017).

Wolfe, C. (2003), *Animal Rites: American Culture, the Discourse of Species and Posthumanist Culture*, Chicago: University of Chicago Press.

Chapter 14

ON OUTCAST WOMEN, DOG LOVE, AND ABJECTION BETWEEN SPECIES

Liz Bowen

> [O]ne of the gifts of genderqueer family making—and animal loving—is the revelation of caretaking as detachable from—and attachable to—any gender, any sentient being.
>
> —Nelson (2015, 72)

> I never understood dog people before I became one, the intense love you can feel for an animal, a gross love that can be without boundaries.
>
> —Zambreno (2015, 102)

Recently, I have been captivated by an emerging thread in contemporary feminist and queer literature: the suggestion that human relationships with domesticated and conventional pets pose a challenge to normative formulations of love and desire. In the epigraph above, for instance, Maggie Nelson reflects on how A.L. Steiner's photo exhibition *Puppies and Babies* represents dogs as sites of interspecies pleasure akin to mother–child intimacies bordering on the taboo. Likewise, in her essay "Notes on a New Tenderness," Kate Zambreno figures human–pet relationships in the base bodily terms they demand but rarely fully engender—think dirt, think saliva, think oral–anal exchange. At the 2017 Carolee Schneemann retrospective at MoMA PS1 in New York, the presence of the artist's cat Kitch is inescapable; the erotic video *Fuses* (1969) appears to adopt the cat's point of view as its human companions have sex. Meanwhile, in popular culture and even outside the mainstream, dogs and cats are regularly deployed as symbols of familial security, figures for nostalgic childhoods, or signals of commitment within socially sanctioned relationships—not terribly energizing conceptual realms. But the alternative possibilities are genuinely invigorating: as anyone who has ever loved a pet knows on some level, *of course* human relationships with pets are gross and intense, and gross in their intensity. Still, these filthy connections have been sanitized in popular discourse, rendered unthreatening to our understandings of species boundedness.[1] The transgressive resonances that Nelson and others locate between human–animal love and queer desire are useful for thinking about human–animal relationality beyond traditional models of ownership, anthropomorphism, and the "unconditional" love we expect from dogs in particular.[2] I can love a dog, and she can love me, but those loves are not identical, unchanging, fully reciprocal, or fully knowable. Not unlike human relationships, companionship between species involves a desirous, delirious, in some ways delusional drive to know one another, in spite of unbridgeable difference. Donna Haraway, another champion of the radical potential of domestic dogs, describes this love as both gross and kind of miraculous: "Significantly other to each other, in specific difference, we signify in the flesh a nasty developmental infection called love. This love is an historical aberration and a naturecultural legacy" (2003, 3).

But dogs aren't simply interesting because they can bring a certain kind of perversity to spheres of heteronormativity and normality, as Nelson suggests. They also bring something complicated to their relationships with humans who are already marginal in other ways. This is visible, for instance, in the love that leads some humans without homes to sleep outside with their dogs even when space is available in a shelter, rather than comply with shelter rules against pets. The interplay of vulnerability and safety is profoundly complex in such a situation, as sticking together exposes both human and dog to violence and the elements at the same time that it provides them with mutual vigilance and protection. To name another example: people with "too many" pets, or the wrong kinds of pets, are typically perceived as perverse or pathological, and their animal attachments may be identified as symptoms of an underlying, stigmatized mental illness. Often, discourses used to denigrate animal lovers are screens for disdain toward other identity categories; stereotypes about people who have too many pets tend to be aimed at figures whose gender, sexuality, age, race, mental health status, and/or class already position them as marginal (e.g., "crazy" old ladies/lesbians or poor people in rural areas).

As a literary scholar, I believe that narrative not only reflects how humans conceptualize their relationships with nonhuman animals but also shapes the form those relationships take—for better and for worse.[3] Thus, to get a better sense of how outsiders and animals relate in human society, I am interested in tracking literary representations of unconventional human–canine love and in exploring the double-edged nature of these literary affinities. In the very same narratives in which these bonds provide enriching affective connections between socially othered humans and dogs, unconfined by heteronormative sentimentality or sanitized domesticity, these relationships can also make their participants vulnerable to violence. Such violence, ostensibly aimed at the human, can be compounded by the additional abjection of the animal, and people seeking to harm humans often end up doing so by harming animals connected with them. Pain, suffering, and precarity are therefore as much a part of these nurturing companionships as pleasure is. In *With Dogs at the Edge of Life*, for instance, Colin Dayan demonstrates how anti-pit bull legislation has been leveraged as a means of policing and oppressing black and low-income white communities in the Southern United States, where both pits and their human companions are judged to be pathologically aggressive (2016, 3–9). It's clear that perceptions of madness and abjection can slip, or be transmitted, between humans and their companion species, often with the result of the animal being subjected to more harm than it would be likely to encounter on its own.[4] So what is at stake, then, when we view these companionships as liberatory? Is there a danger of eliding pain and precarity, akin to the way feminist reclamations of the "madwoman" obscure actual experiences (and difficulties) of mental illness?[5] What might become invisible when we celebrate the transgressive potential of interspecies love? In asking these questions, I do not wish to evacuate the radical potential offered by against-the-grain readings of human–animal love, but to offer a fuller account of their risks *and* possibilities.

I will do so with the help of two narratives that dramatize unusual affiliations between socially outcast humans and their dogs: *Disgrace* by J.M. Coetzee (1999) and Alice Notley's *Culture of One* (2011), described by the publisher as a "novel in poems." Both texts tell a story of a woman living at the outer edges of society, arguably in closer relationship to dogs than to other humans, in ways that others deem mad, irrational, and/or queer. Both also depict violence against these women as closely bound with violence against their dogs, as the characters are forced to witness the murder of their animal companions as punishment for their own social transgressions. In *Culture of One*, the central character Marie, who lives on a landfill with her dogs in the American Southwest, endures a series of humiliations at the hands of people more powerful than she is: her shack is burned down by a male sexual partner, killing her infant daughter, and a group of acrimonious teenage girls murder her beloved coyote-mutt Tawny. Meanwhile, Coetzee's Lucy Lurie, a white South African lesbian who prefers a rural life boarding dogs to the more urban proclivities of her young-adult generation, both solidifies and complicates the connection between dogs and social exclusion. While Lucy's queerness and passion for animals render her an outcast and thus a particularly visible target for violence, this outsider status also makes her a conspicuous representative of the history of white domination in her community. Specifically, the tradition of white South Africans training dogs to attack black people is suggested to be at least one part of what makes Lucy a target in the recently post-apartheid moment of the narrative. Although there are important differences between Marie and Lucy and between their attackers, both texts foreground human–animal love as a marker of outcast status, as a site of both nurturing and traumatic encounters, and as a justification for certain kinds of animals—human and nonhuman—being rendered repellant, disposable, and unworthy of compassion.

This transference of abjection from woman to animal destabilizes some of the taxonomies of animal life that we have come to take for granted in animal studies, such as the distinction Cary Wolfe makes between killable-but-not-murderable livestock and domestic animals like dogs—which, as members of communities, are supposedly considered murderable and grievable.[6] But if dogs are members of the wrong communities, these narratives suggest, they become simultaneously killable and murderable: killable to those who kill them and murderable to the humans left to grieve them. This is a devastating outcome, more in line with Derrida's characterization of the nonhuman as that which is always left open to sacrifice, or a "noncriminal putting to death" (1991, 112). However, I will argue that these novels' formal preoccupations with how to narrate experiences of trauma, depression, and grief, which resist incorporation into conventional or realist narrative forms (i.e., linearity, univocality, closure), allow readers to hold together their depictions of the tragic and redemptive possibilities of human–canine affiliations. Moreover, these texts' interpolation of violence against animals into systems of human oppression illuminates how ethical and political questions typically assumed to be limited to the human realm—that is, race, class, ability, and gender—are founded on a privileged concept of the human whose limits put both humans and nonhumans at risk. In other words, to quote Judith Butler, "nonhuman life is also precarious life and ... precariousness links human and non-human life in ethically significant ways" (Antonello and Farneti 2009, n. p.). Without minimizing the loss of animal life or sentimentalizing its moral import, these novels allow us to imagine futures that could emerge out of experiences at the edges of normative rationality and the human, in which a kind of becoming-doglike must be the condition for ethical action.

Disgrace

> They are not going to lead me to a higher life, and the reason is, there is no higher life. This is the only life there is. Which we share with animals. —Lucy Lurie (Coetzee 1999, 74)

When Professor David Lurie, the unsympathetic protagonist in J.M. Coetzee's *Disgrace*, arrives at his daughter Lucy's rural home, he is none too pleased with her secluded lifestyle and atypical gender expression, but he does not find it his place to tell her so. David, whose recent affair with—and rape of—a student has landed him in the midst of a career-ending public scandal, arrives at Lucy's farm grumbling to himself about her choice to defect from her former human community to live among the dogs she shelters. These animals are "watchdogs," which Marianne DeKoven notes is code for "temporarily out-of-service guard dogs: agents of the enforcement of apartheid whose services are now only sporadically required" (2009, 850). While Lucy's harboring of these animals signals her racial privilege and her complicity in white supremacy in South Africa, it also illuminates her vulnerability as a queer woman living far away from any queer community. However, though the novel's free indirect style sticks closely to David's condescending point of view, it quickly becomes clear to the reader that Lucy has cultivated an existence that is satisfying to her in its austerity and alterity, and that this satisfaction is predicated on her love for dogs. Even David sees how her animal love fits into a constellation of subject-forming affinities: "Now, in her middle twenties, she had begun to separate. The dogs, the gardening, the astrology books, the asexual clothes: in each [David] recognizes a statement of independence, considered, purposeful. The turn away from men too. Making her own life" (89). While David interprets Lucy's non-normative interests as a *separation* from the status quo, her affection for animals encourages us to read these signs as indicative of a deeply *relational* existence—that of a social outcast who nonetheless recognizes potential for affiliations "outside the bounds of civil life" (Dayan 2011, 232). When David remarks glibly on his romantic life, "desire is a burden we could well do without," Lucy responds in earnest, "I must say … that is a view I incline toward myself" (Coetzee 1999, 90). And yet it is clear that her desire for relationships with nonhuman animals is expansive and expanding: since David's last visit, she has increased her kennel capacity fivefold and has plans to begin taking in cats. Lucy's impulse to share her world with animals is ethical and visceral, as well as unbounded by the notions of ownership and unconditional love that Haraway so acerbically names "the neurosis of caninophiliac narcissism" (2003, 33).

But this is not where the story ends. In a jarring turn nearly halfway through the novel, a group of robbers force their way into Lucy's home, rape her, and massacre the dogs in the kennel. In the aftermath of this attack, Lucy experiences a major depressive episode defined by silence and passivity, which bewilders and frustrates her father. She never explicitly articulates her reasons for choosing not to narrate her rape to authorities, and so David fixates on the racial identity of her attackers, assuming that since they were a group of black Africans, Lucy is forgoing justice for herself to atone for South Africa's history of racist violence.[7] David misreads Lucy's behavior in much the same way that, earlier in the novel, he misreads her compassion for animals as a matter of guilt or "fear of retribution" (74). Her earlier professed desire not to "come back in another existence as a dog or a pig and have to live as dogs or pigs live under us" (74) is not a matter of guilt, but of sympathetic identification; she can imagine the life-world of domesticated animals and wishes to build a less traumatic alternative. When viewed in the context of her actual relationship to the dogs, then, her depressive reticence following her own traumatic event becomes more comprehensible than David makes it out to be: Lucy responds to the violence committed against her as a dog would. As Dayan so beautifully observes, dogs who die at the hands of human violence appear to exhibit a similar sort of detached passivity: "When dogs find themselves in the wrong place at the wrong time, belonging to the wrong kinds of people or protecting too earnestly the homes of their human companions, they gather themselves up in their flesh, and in a state of prescience and acceptance, they prepare for the time when life stops, as they slip away toward stillness" (2016, 315).[8] This is precisely how Coetzee depicts the deaths of Lucy's dogs while they witness the killing of their kennelmates. As the gunman picks off the animals with his rifle, a dog with a "gaping throat-wound" "sits down heavily, flattens its ears, following with its gaze the movements of this being who does not even bother to administer a *coup de grâce*" (95). Then, a "hush falls," and the few dogs left alive retreat to the back of their pen until they, too, are killed (96). For a person committed to living with and learning from dogs, Lucy's resignation as she declares "I am a dead person" to her father is more than just a representation of her own mental distress; it is an expression of an interspecies, doglike disposition in the face of human violence (161). While Lucy's turn away from society and toward identification with her politically incorrect dogs is certainly suspect in a post-apartheid economy of social engagement and reconstruction, it is important to note that the dogs, too, are political subjects. They have been conscripted agents of the apartheid regime, relegated to a life of confinement, occasional work, and vengeful death. Thus, when Lucy responds to her attackers' violence *like a dog*—adopting a state of inanimacy that, to mainstream society (as represented by her father), seems irrational and nonhuman[9]—Lucy takes on the affect of the only characters in the novel who face mortal consequences for their complicity in that regime.

Though much has been written about the importance of sympathetic imagination in *Disgrace*, this turn in the novel also suggests an ethics beyond just imagining oneself in an other's position.[10] After the attack, as David Lurie comes to grips with the fact that he cannot access his daughter's traumatic experience from his position of logic and reason, he develops a number of close relationships with animals. As Laura Wright has written, "[David's] sensibility with regard to the lives of animals is greatly altered after Lucy is raped, and this change in perception results from a respect for the being of animals regardless of whether or not David can access the animal's interiority" (2007, 92).[11] Simply valuing others' lives without being able to identify or imagine one's way into them is not itself a radical political stance—let's not forget, white supremacists could love their attack dogs, too—but the capacity to do so, which David Lurie surely does not possess at the beginning of the novel, is a necessary basis for any ethical action on the part of a dominant group. The passivity of Lucy's depression and retreat from narrative language, both of which are wholly inaccessible to her father, are what bring him to a place where he might cease reproducing the violence he has wrought in the past precisely by seeking total access to others' lives and bodies.

We might ask, however, if this is where a story of dog love like Lucy's has to end up: death and devastation in service of the questionable ethical development of the white male rapist who remains at the story's center. This outcome is frustrating, to say the very least. Still, I maintain that the outcast status of women in this text puts crucial pressure on David's centrality to the novel. This is evident in the character of Bev Shaw, a friend of Lucy's whom David finds utterly repulsive. He remarks repeatedly on her unkempt, unshapely appearance and overweight body, and when he enters her home, he is "repelled by the odours of cat urine and dog mange and Jeyes Fluid" (72). Bev takes a moralistic view toward her job euthanizing dogs at the local clinic, suggesting that if someone has to do it, it's better that it be someone who is deeply bothered by it. With Lucy's story in mind, we might read her philosophy as a self-defense against the tragedy and vulnerability that human–dog love makes possible. Bev recognizes that she lives in a society that slaughters animals due to their proximity to humans, and rather than open herself up to that same unpredictable vulnerability, she short-circuits it by undertaking the killing herself. Lucy's response enacts a different kind of ethics to respond to the same problem of animals' entanglement in human sociopolitical systems. In living with depression visibly and unapologetically, becoming abject, and responding to violence like a dog, she refuses the values of human superiority wholesale, a refusal she suggests is necessary to combat all kinds of human-perpetrated violence. Cast against these women, David, who only at the end of the novel has begun to ponder whether a dog might have a place in the human tragedy of the opera he is writing, ends up looking rather like an oblivious footnote to the theory of interspecies oppression that the outcast minor characters have been putting into practice throughout.

Abjection

Why don't you know how to talk to people? I talk to the dog-masks of the dogs … Your eye is mild in the swirls of your dog-mask, its gentle fibers: your eye is mild. Your mask is used to remove the taboo against two species connecting with each other.— Marie (Notley 2011, 25)

Alice Notley's *Culture of One* is, like *Disgrace*, a brutal endeavor. Its central character is rendered inhumanly abject on a routine basis; as an antisocial, aging woman who lives on a landfill at the outskirts of a desert town, she is regularly tormented by hate-spewing teenagers and abused by the few people she maintains relationships with. Her distance from human society is suggested to be partly voluntary, which earns her a reputation for being "crazy"—a madwoman whose love for her dogs is so intense she is rumored to eat them. Like Lucy, Marie does not tend to express dissatisfaction with her isolation from humans, but rather satisfaction with her sense of belonging among dogs. Reflecting on a group of "cruel chicks" that shows up at the dump to taunt her, Marie casts them as inferior to her dogs: "I'm trying to remember; the girls were like my dogs. They liked/suddenly to growl: you have the wrong mask you won't be loved./ … I hate you the girl says. You're not like my dogs I say,/you're constructed: glued-on hair and fangs and pulpy genitals" (17). She suggests that although these girls' doglike behaviors show their potential for interspecies connection, they are impoverished in their inability to recognize their own animality the way she can. Their aggressive behavior does not animalize them, but rather exposes their cruelty as a mask, a construction that renders them ontologically subordinate to animals: "just humans; they don't even look/alike, like iguanas, or pallid bats, or pack rats—perfect/animals" (97).

Importantly, though, both Marie and the poems themselves refuse a simple dichotomy in which animals equal natural and good, and the human is cultural, constructed, and therefore corrupted. The "culture of one" to which the book's title refers is, in fact, the "new/way of living" that Marie constructs for herself and her dogs in the dump, hoarding thrown-away scraps to build her home and an impressive portfolio of artworks (60). She spends her days working on a project called the codex, an elaborate alphabet book created out of found objects and media from the dump, encoding each page with associations and materials particular to her experience. The pages' meanings are esoteric—clear only to her, the narrator, and sometimes the reader (e.g., D stands for dog, R for the name of the man who killed her child). But as one of the text's primary narrators—a mysterious goddess of Mercy loosely based on the Tibetan Bodhisattva Tara—tells us, culture "comes from the materials you do it with" (10). Marie's pursuit of an individual culture is both a repudiation of mass culture and a reimagining of its possibilities, participating in it at the level of its discarded materials. She also considers the dogs' vivacity an affirming force for her aesthetic vision and work, noting that their embodied traces constitute the very materials of her project:

> The dogs tell me it could be worse: I could be eaten by oversubtlety rather than bold red or blue letters howling. HOWLING! That word again. The dogs open their mouths to word me.
>
> Have you ever dreamed you didn't have a master. We dreamed we ran down the gully to the river, but not without you; we couldn't leave you … Everything's covered with dog hairs; shake them off the illuminated pages, no they're painted on, ocher, gold, and black filaments. (88)

Marie's cultural products are not really "of one"—they are an archive of a life lived among stray animals that, like her work's other materials, are considered extraneous to the town's domestic and economic systems. Marie is, however, what we might call an outsider artist: self-taught, unengaged with the genealogy and traditions of "high art," and excluded from the capital-driven, star-powered art world. The "outsider" genre label has come under well-deserved scrutiny for its encouragement of ableist assumptions that non-neurotypical people lack cultural influences, its presumptive interpretations of artworks as literal representations of the artist's "inner world," and its reification of marginalized artists' status as socially and economically other.[12] Still, there is one commonly cited aspect of outsider art that seems germane to Marie's role as an artist: a presiding interest in process rather than product. Like the sculptor Simon Rodia, who spent thirty years constructing a complex of large-scale structures and then permanently walked away from them, or Henry Darger, who wrote a 15,000-page illustrated epic and then told his neighbor to throw it all away, Marie is devoid of sentimental attachment or preciousness about the products of her labor. When it comes to the precarity of her artworks, she echoes outsider artist The Pope of Montreal, who, when his massive installation of hats was destroyed in a fire, said only, "Well, that's sad, but I will do it again" (Boxer 2013). Note Marie's almost identical response to losing her work in a fire: when "Every once in a while a kid burned down her shack …. She'd start again. She always remembered how to do it" (10). When asked, "What are you going to do when they burn up your shack?" Marie replies, "I don't care, it'll still be great here" (10). What makes Marie's home great is not that it houses her work, but that it provides the materials for its creation and recreation. The process of repurposing the detritus of a culture that considers Marie and her dogs to be waste constitutes a practice of resistance against the toxic systems of production, consumption, and valuation that locate them there.

Though Marie is not diagnosed with any disability, it is clear that she is alienated from artistic and intellectual discourse, including the poetics of the text that narrates her life. The novel-in-poems repeatedly notes that it is using words to describe Marie that she would not know herself, including the word "culture." In this way, to borrow a term from disability studies scholar James Berger, Marie is disarticulate(d) not just from her fictional community but also from the narrative voice and aesthetic sensibility that constructs her. Berger defines the disarticulate as "the figure at the boundary of the social-symbolic order, or who is imagined to be there, and at that liminal place, there is no adequate terminology. One cannot even quite determine whether he is an object of desire or revulsion" (2014, 1). Like this figure, Marie is rendered abject by her status as a linguistic other, making her an unstable body that both fascinates and repels the people around her—as well as the text itself, which attempts to represent her at the same time it avows its failure to do so. While the cruel chicks find Marie repellant to the point of incredible violence, Notley's narrative voice claims that a lifelong fascination with Marie has led her to write a book that betrays her: "I mean *everyone* lies about Marie. I'm lying about her right now,/ in order to pay my respects," she says (21). Notley's commentary on her own betrayal signals an uneasy affiliation with Marie's antagonists. The cruel chicks and the narrator, though seemingly opposed in their attitudes, represent an interplay of desire and revulsion that renders Marie both heroic and degraded, or "beside herself," as Julia Kristeva describes the state of abjection (1982, 1). Whether Marie is portrayed as beside herself as a literary figure—an idealized misrepresentation of an actual person outside the socio-symbolic order—or whether she's beside herself with grief as a result of other characters' violence against her, she is placed in an abject position both by the traumatic events of the novel and by her very existence as a figure within it. Unlike Marie's artwork, the aesthetic tools available to Notley's linguistic medium can only approximate and appropriate Marie's experience, with self-consciously uncertain ethical implications.

Still, the text's attempt to capture in language Marie's eccentric and extralinguistic modes of communication with her dogs enables the reader to imagine what Marie's codex, which she repeatedly claims is her brain and her world, might look and feel like. Though the codex is an alphabet book, Marie's artmaking process is depicted as less akin to writing than to howling, an opening of the body "to word" without or beyond language (i.e., "red and blue letters howling … The dogs open their mouths/to word me"). The letters in her alphabet do not follow a conventional order, but rather correspond to Marie's emotions and experiences at the moment she creates them. Though the letters do sometimes stand for meaningful words in Marie's life, they also often become figures in their own right, defined more by color, materials, or affective charge than by any correspondence to language. The red and blue letters howl not with the "oversubtlety" of poetry, after all, but with the memory and materials of the dump. It is no accident that this process of constructing a text that does not conform to standardized lexicons is figured in animalized terms, since membership in a linguistic community is one of the primary abilities that supposedly separates humans from other animals. As Sunaura Taylor has pointed out, this is misconception that relies on ableist assumptions about human communication, and both humans and nonhuman animals voice their discontent and suffering in forms that proceed and exceed language (2017, 43). Like the pained lowing of a dairy cow whose calves have been taken away by farmers, howling for Marie has a communicative function, if not always a semiotic one. The codex, in its perpetual state of reconstruction, is its own traumatic return, a materialization of the unspeakable experience of being tortured, or being stripped to bare animality in the face of human cruelty. Elaine Scarry writes that witnessing a person in great pain transition from prelingual cries and groans to fragmented verbal expressions is "almost to have been permitted to be present at the birth of language itself" (1985, 6). Marie's reconstitution of the alphabet in extralinguistic form might well be understood to occupy just such a state of creation.

The world that Marie builds in the codex not only disregards conventional syntactical and semiotic forms, but even reaches further than the rather experimental poetic form in which Notley tells Marie's story. In true Notleyan fashion, the voice of *Culture of One* is never just one voice, as the narratorial position slips between the Goddess of Mercy, Marie, and a Southwestern expat living in Paris who represents Notley herself. While all those voices are quite similar syntactically and stylistically—are all, in a sense, Notley's voice—they are all centered around the codex, intent on including its strange alphabet despite its communicative otherness. As the Notley character agonizes over the cultural embeddedness of her own language, which she wryly bemoans as "Popup thoughts for scholars," the alphabet of Marie's codex builds a language utterly estranged from human community and interpretability, in symbols quite literally made up of dog DNA ("everything's covered with dog hairs … they're painted on") (129, 88). The codex constructs a grammar of abjection that can accommodate Marie as a nomad to society and language. This is quite different from Lucy Lurie's linguistic otherness, which is defined by depressive silence and refusal, the near opposite of Marie's "crazy" howling and hoarding

(26). Nonetheless, both are able to imagine and instantiate new worlds through linguistic practices traditionally deemed subhuman or pathological.

The problem for Marie and her dogs, though, is that very sense of shared and shifting territory that empowers her to be an artist on the outer edges of culture: in running down the gully to the river together, untethered and unashamed, she and her dogs provoke the disdain of the nearby community. Others can't bear to see her claim territory *outside*, whether outside their standards for what constitutes a human life (a life neatly distinguished from nonhuman life) or outside their human standards of living (a life at a sane and sanitary distance from the dump). The diagnosis of insanity based on transgressed boundaries, whether they are city limits or the number of dogs a person should own, are precisely the kinds of territorial disputes that Rosi Braidotti argues form the foundations of human–animal relationality. In *Nomad Theory*, she writes that becoming-animal requires a becoming "radically immanent" to one's environment, and Marie's willingness to incorporate and be incorporated into her extrahuman culture is what makes both her and those with whom she shares that culture a threat (2012, 94).

Much like Lucy Lurie, Marie finds that her nonnormative life with dogs exposes both her and the dogs to gender-based violence, which is ostensibly directed at the human but ultimately fatal for the dogs.[13] The poem's speaker tells us: "Marie is the truth. Treating her badly—an older/woman—is the world's delight. Her search for a new way/to live, her determination to create a cell of beauty/and meaning around her—despite being in a dump ... /We hate her: she's a witch" (139). While Marie is demonstrably traumatized by each act of violence against her, she also seems convinced of the inescapability of that violence, as if it is an uncontrollable element of her habitat. After a group of teenage girls feeds broken glass to Tawny, who dies in Marie's arms, Marie displays rage more visibly than Lucy Lurie, and yet registers the futility of her grief-fueled animacy. While berating the girls, she repeatedly reflects on how "banal" and ineffectual her response is: "Marie turns and says, You're/just anyone, and that's the worst thing I can say. I'm/being so banal, she thinks, without thinking the word./ Because, people hurt you and make you confront them—/so banal" (109). Tawny's death is, it seems, the logical, even inevitable consequence of a world in which people hurt people like Marie, so-called crazy women who build their homes at the edges of the human. And yet, still, she carries on with her life among dogs: caring for them, running with them, spending whatever money she has on them—and, yes, sometimes, burying them.

Affiliation

Again, I find myself asking: Is this how these stories have to end? Is it the case that dogs are so already-abject that we can only imagine their association with vulnerable humans ending in their deaths? If so, are we obscuring animals' pain by celebrating interspecies connections as sites of desirous potential, aesthetic innovation and alternative subject formation? Are these meaningful, affectively vibrant, and theoretically challenging relationships worth the risk they pose to animals' lives (as well as to the humans who must mourn them)?

These are difficult questions, and in the world outside novels, the interplay of enriching companionship and shared precarity in human–dog partnerships is contingent on the specific circumstances of particular humans and their animals. In the context of fiction, though, I would suggest that there is more room for reparative potential when it comes to the deaths of companion species. The representation of dogs' dual status as "[b]earing the brunt of the most extreme denigration and at the same time enjoying the greatest affection" encourages reflection on how ostensibly human structures of oppression have ethical consequences beyond the human (Dayan 2011, 231). Narratives like *Disgrace* and *Culture of One* demonstrate how human relationships to another species never can be unconditional; the dog's status as denigrated or beloved is entirely dependent upon their position within social hierarchies of race, gender, class, sexuality, and ability/normality. By dramatizing this duality, these narratives demand a reinterrogation of "the human" as a privileged and bounded category.

In the meantime, they suggest, there is perhaps no better option for beings on society's margins than affiliation, even as it might render both humans and animals more vulnerable. Take, for instance, the following passages from the ends of our two texts:

> (Lucy to her father, after deciding to bear a child conceived during her rape:)
> "Yes, I agree, it is humiliating. But perhaps that is a good point to start from again. Perhaps that is what I must learn to accept. To start at ground level. With nothing. Not with nothing but. With nothing. No cards, no weapons, no property, no rights, no dignity."
> "Like a dog."
> "Yes, like a dog." (Coetzee, 205)

> (Notley's narrator:)
> Mercy left her concept in place; perhaps
> I'll tear it down. Who deserves Mercy except for Tawny,
> Marie, or Leroy, or the Satanist girl or maybe Eve?
> I deserved it last year. Because this book is *my* world,
> and Mercy still knows my friends. My friends at the dump. (Notley, 121)

Whereas the kind of human–animal brutality depicted in these narratives may leave room for little but trauma in the real world, the imaginative capaciousness of the novel allows us to conceive of more merciful futures that emerge from trauma but are not confined to it. *Disgrace* is in many ways a pessimistic novel,[14] but I argue that Lucy's becoming-doglike suggests the possibility of undermining a social and narrative order defined by the brutality of human hierarchy. As Dayan asks, "In a country wracked by economic collapse, ingrained racism, and political paralysis, where else is there to go, except into the eyes of dogs?" (Dayan 2015, 121). That is, in a society where an end to humans' legacy of racial disgrace based on dehumanization cannot be easily imagined, a radical deprivileging of the human may be the only hope for escape. Meanwhile, Notley envisions the murder of Tawny as an invocation to mercy for her abject and irrational subjects, a mercy she is only able to generate through narration. Mercy, a disenchanted goddess who has gone into retirement rather than continue to witness "mercy" being wielded by sovereigns and dictators, is called back into action by Tawny's death: "Mercy despised her job: But the agony of a half-coyote/dog with glass in her guts hurts like hell .... We must redefine mercy now, [she says]; it can't/belong as a concept to leaders and exploiters" (107–121). By constructing a world in which mercy is granted *primarily* to beings like Marie and her dogs, Notley both renders such a world imaginable and demands its materialization. Despite the cruelty of their plots, these are narratives that love dogs and the women who love them too much—not in the sentimentalized modes of pet ownership, but with clear-eyed concern for the ways animals' intimacies with and differences from humans put them at risk. When love between dogs and humans can be so meaningful, so full of potential for new imaginings, it is unlikely that these interspecies affiliations will cease. And so, these novels suggest, in a world where such exquisite encounters are swallowed by the machinations of human power and violence, perhaps the best way forward is to become more like dogs.

Notes

1   For more on how love between domesticated animals and humans "transcend[s] the boundaries of their bodies and their species," see Kathy Rudy (2011, xii).

2   Donna Haraway also rightly criticizes this "unconditional love" (2003, 38).

3   Marianne Scholtmeijer makes this point about outsiders and animals in "The Power of Otherness: Animals in Women's Fiction" (1995, 233).

4   Dayan notes that while both humans and dogs are harmed by breed bans, it is ultimately the animals whose lives are at stake due to compulsory euthanasia (2006, 10).

5   See Elizabeth Donaldson (2002, 11–31).

6   Cary Wolfe writes, "[M]any animals flourish ... because they are felt to be members of our families and our communities, regardless of their species. And yet, at the very same moment, billions of animals in factory farms.... Clearly, then, the question here is not simply of the 'animal' as the abjected other of the 'human' tout court" (2012, 53).

7   For more on the political stakes of this misreading, see Rosemary Jolly (2007, 164).

8   While this is certainly an anthropomorphic reading, much critical animal studies scholarship admits a certain level of practical anthropomorphizing in service of

developing a human–animal ethics. See Vinciane Despret (2004, 111–134); Cora Diamond (1978, 465–479); Jane Bennett (2010); and Tom Tyler (2003, 267–281).

9 The concept of inanimacy is borrowed from Mel Y. Chen's engagement with animacy as a linguistic and cultural construct. See Chen (2012, 1–20).

10 For more on Coetzee's interest in the sympathetic imagination, see Sam Durrant (2006, 118–134); and Michael Marais (2006, 75–93).

11 Derek Attridge puts it another way: "If Lucy and Petrus are for David Lurie others whom he struggles to know, if [his victim] Melanie is an other whom he wrongs by not attempting to know, animals are others whom he cannot begin to know" (2000, 113).

12 See David Mclagan (2009, 35–41) and Benjamin Fraser (2010, 513).

13 It is important to note the major difference between the acts of violence in these two books, which is that there is no racial aspect of *Culture of One* akin to the post-apartheid race relations that play a role in Lucy's rape. The class-based violence against Marie is not analogous to the racial tensions in *Disgrace*. For more on the danger of making such analogies, see Stefanie Boese (2016, 249).

14 For more on its pessimism, see Mark Sanders (2011, 363–373).

References

Antonello, P. and Farneti, R. (2009), "Antigone's Claim: A Conversation with Judith Butler," *Theory & Event* 12 (1). Available online: https://muse.jhu.edu/article/263144 (accessed December 13, 2015).

Attridge, D. (2000), "Age of Bronze, State of Grace: Music and Dogs in Coetzee's *Disgrace*," *NOVEL: A Forum on Fiction* 34 (1): 98–121.

Bennett, J. (2010), *Vibrant Matter: A Political Ecology of Things*, Durham: Duke University Press.

Berger, J. (2014), *The Disarticulate: Language, Disability, and the Narratives of Modernity*, New York: New York University Pres.

Boese, S. (2016), "J.M. Coetzee's *Disgrace* and the Temporality of Injury," *Critique: Studies in Contemporary Fiction* 58 (3): 248–257.

Boxer, S. (2013), "The Rise of Self-Taught Artists: Out Is the New In," *The Atlantic*, September. Available online: https://www.theatlantic.com/magazine/archive/2013/09/out-is-the-new-in/309428/ (accessed July 8, 2017).

Braidotti, R. (2012), *Nomadic Theory*, New York: Columbia University Press.

Chen, M. Y. (2012), *Animacies: Biopolitics, Racial Mattering, and Queer Affect*, Durham: Duke University Press.

Coetzee, J. M. (1999), *Disgrace*, New York: Penguin.

Dayan, C. (2011), *The Law Is a White Dog: How Legal Rituals Make and Unmake Persons*, Princeton, NJ: Princeton University Press.

Dayan, C. (2015), *With Dogs at the Edge of Life*, New York: Columbia University Press.

DeKoven, M. (2009), "Going to the Dogs in *Disgrace*," *ELH* 76 (4): 847–875.

Derrida, J. (1991), "Eating Well, or the Calculation of the Subject: An Interview with Jacques Derrida," in E. Cadava, P. Connor, and J.-L. Nancy (eds.), *Who Comes after the Subject?* 96–119, New York: Routledge.

Despret, V. (2004), "The Body We Care For: Figures of Anthropo-Zoo-Genesis," *Body & Society* 10 (2–3): 111–134.

Diamond, C. (1978), "Eating Meat and Eating People," *Philosophy* 53 (206): 465–479.

Donaldson, E. (2002), "The Corpus of the Madwoman: Toward a Feminist Disability Studies Theory of Embodiment and Mental Illness," in D. Bolt, J. Miele Rodas, and E. Donaldson (eds.), *The Madwoman and the Blindman: Jane Eyre, Discourse, and Disability*, 11–31, Columbus: Ohio State University Press.

Durrant, S. (2006), "J. M. Coetzee, Elizabeth Costello, and the Limits of the Sympathetic Imagination," in J. Poyner (ed.), *J. M. Coetzee and the Idea of the Public Intellectual*, 118–134. Athens: Ohio University Press.

Fraser, B. (2010), "The Work of (Creating) Art: Judith Scott's Fiber Art, Lola Barrera and Iñaki Peñafiel's *Qué tienes debajo del sombrero?* (2006) and the Challenges Faced by People with Developmental Disabilities," *Cultural Studies* 24 (4): 508–532.

Haraway, D. (2003), *The Companion Species Manifesto: Dogs, People, and Significant Otherness*, Chicago: Prickly Paradigm Press.

Jolly, R. (2007), "Going to the Dogs: Humanity in J. M. Coetzee's *Disgrace, The Lives of Animals*, and South Africa's Truth and Reconciliation Commission," in J. Poyner (ed.), *J. M. Coetzee and the Idea of the Public Intellectual*, 148–171, Athens: Ohio University Press.

Kristeva, J. (1982), *Powers of Horror*, New York: Columbia University Press.

Marais, M. (2006), "J.M. Coetzee's *Disgrace* and the Task of the Imagination," *Journal of Modern Literature* 29 (2): 75–93.

Mclagan, D. (2009), *Outsider Art: From the Margins to the Marketplace*, London: Reaktion Books.

Nelson, M. (2015), *The Argonauts*, Minneapolis: Graywolf.

Notley, A. (2011), *Culture of One*, New York: Penguin.

Rudy, K. (2011), *Loving Animals: Toward a New Animal Advocacy*, Minneapolis: University of Minnesota Press.

Sanders, M. (2011), "Disgrace," *Interventions: International Journal of Postcolonial Studies* 4 (3): 363–373.

Scarry, E. (1985), *The Body in Pain: The Making and Unmaking of the World*, Oxford: Oxford University Press.

Scholtmeijer, M. (1995), "The Power of Otherness: Animals in Women's Fiction," in C. Adams and J. Donovan (eds.), *Animals and Women: Feminist Theoretical Explorations*, 231–262, Durham: Duke University Press.

Taylor, S. (2017), *Beasts of Burden: Animal and Disability Liberation*, New York: The New Press.

Tyler, T. (2003), "If Horses Had Hands …," *Society & Animals* 11 (3): 267–281.

Wolfe, C. (2012), *Before the Law: Humans and Other Animals in a Biopolitical Frame*, Chicago: University of Chicago Press.

Wright, L. (2007), "'Does He Have It in Him to Be the Woman?': The Performance of Displacement in J.M. Coetzee's *Disgrace*," *Ariel* 37 (4): 83–102.

Zambreno, K. (2015), "Notes on a New Tenderness," *Animal Shelter* 4: 101–111.

DISCUSSION BY CAROL J. ADAMS

Oh animaladies—*Animaladies,* the book and *animal-ladies,* the authors—how I do love you! How you open up and restore ways of being. My love comes with sadness, grief, a sense of so much that has been lost, of damage done, of opportunities missed. All the possibilities of beingness that have been lost to us—constrained and labeled and killed. Over and over again, the pathologizing of interspecies connections, especially those that women have made.

Disturbance

Can I say it is a mad love? You have aroused my anger. Do I rage? Yes I rage. I think about Linda Nochlin, the great feminist art critic, writing about rage: "It seems to me, then, important to examine not merely how rage might be said to get into painting or sculpture but also how it gets into women" (2002, 51). She considers gendered rage in writing about Joan Mitchell, an abstract expressionist painter, and quotes from a letter Mitchell sent to her in the late 1980s,

> I have lots of real reasons to hate ... and somehow I can't ever get to hatred unless someone is kicking my dog Marion (true story) ... [*sic*] or destroying Gisèlle [a friend] etc. and then I'll bite—(I can't get to killing—my 'dead' shrink kept trying to get me there—I have never made it—my how I loved her). (52)

How much *animaladies* that excerpt contains.

When Nochlin goes on to write about how "meaning and emotional intensity are produced structurally" (55), I think of lynn mowson's *boobscapes* and *slink*. I rage about the slaughter of pregnant cows, and am grateful for lynn and Katie Gillespie for finding receptacles by which to mediate such rage—reflections for Katie, sculptures for lynn.

lynn refers to my own use of the term "traumatic knowledge." I learned that term from feminist historian Bonnie Smith's *The Gender of History.* Smith describes how amateur history (the history associated with women) consists of "the writing of multiple traumas" (2000, 19). She identifies the traumas that women historians of the early nineteenth century would have experienced: They were aware that their rights were eroding in the midst of a time when universal rights were being (supposedly) championed. They or family members had survived revolutions and wars, and at a personal level had experienced the threat or actuality of rape, poverty, violence, abuse. I applied this concept to the experience of those who care about animals and the knowledge we acquire about their experiences.

An important part of our response is to share and reveal; we seek ways to be creative. We write. Traumatic knowledge often leads to empathic witnessing, but I sense there is rage too.

Distraction

My love is mad, angry, destabilizing.

This morning I have been remarkably efficient in anything that took me away from this book—looking up a recipe for creamy vegan grits (searching five cookbooks for the best one). Thinking of the fact that *two* of my great aunts (eight times removed) were killed as witches in Salem, Massachusetts, in the 1690s. This matters. I stood in front of my books on witchcraft ("mad" women) and pulled out Keith Thomas. Many witches were said to have "witches' familiars," that is, companion animals said to be equally bewitched (especially European women accused of being witches).

As Keith Thomas describes it:

> But whether these domestic pets or uninvited animal companions were seen as magical is another matter. These creatures may have been the only friend these lonely old women possessed, and the names they gave them suggest an affectionate relationship. Matthew Hopkin's victims in Essex included Mary Hockett, who was accused of entertaining "three evil spirits each in the likeness of a mouse, called 'Littleman,' 'Prettyman' and 'Daynty,'" and Bridget Mayers, who entertained "an evil spirit in the likeness of a mouse called 'Prickeares.' More recently the novelist J. R. Ackerley has written of his mother that: "One of her last friends, when she was losing her faculties, was a fly, which I never saw but which she talked about a good deal and also talked to. With large melancholy yellow eyes and long lashes it inhabited the bathroom; she made a little joke of it but was serious enough to take in crumbs of bread every morning to feed it, scattering them along the wooden rim of the bath as she lay in it." (1971, 525)

It is a good world when a fly is fed.

Discomfort

It is hard to encounter these essays; they ask a lot of us and we discover that we have that much to give.

I needed comfort food; that explains my search for a good recipe for grits.

These essays offer us a way of understanding. Giving us liberation into *being.* Liberation into grief (Thank you Vas and James Stanescu).

For comfort food, I tried slow cooked lima beans (already made) as a substitute for the grits (which were unmade).

Disembodied

If only embodiment were sufficient. We know of the rationalized definitions that defined "human" by disregarding the body, even though thinking is a bodily function.

How we are supposed to survive, if we practice disembodiment? The *bodied*iment that Katie encounters is important, as is her witnessing—truly an act of embodying, of being painfully embodied. Yet how much of embodiment is unintelligible to others and their disembodiment is normalized. I like lynn's thought: "that one might empathically respond to the materiality and form, and in doing so recognize these things we share with our fellow creatures, our entangled embodiment." And I recall Hannah Monroe and her reminder of the overly rationalist way of discussing animal liberation that has privileged mind over body.

Disaffection

Disembodiment is connected to dis/affection: the way affection and emotion in general have been stigmatized, misunderstood, dismissed. Lori Gruen asks how can we be fully in meaningful relationships when an important part of our cognitive/affective capacity is cut off? How is the definition of human already a dis/abling discourse? Hannah reminds us, "Autistic people are constructed by medical discourse as being unable to empathize and yet empathy is constructed as solely a feminine trait." Who bears the burden and liberation of caring too much (who says "too much"?) and wrong kinds of caring (again, who says?). Heather Fraser and Nik Taylor explore how the "nervous disposition" (and the stereotypes it bears) manages anxiety with the help of animal companions, but how this understanding pathologizes women's emotions for animals.

Fiona Probyn-Rapsey looks at "caring gone wrong" and how it too pathologizes individual acts of collecting animals but ignores the issue of whether factory farmers aren't themselves hoarders. Who is a hoarder? Why are women pathologized, but not factory farms? What is "too far" and according to whom? Is it because an individual hoarder is involved in a process but factory farms are committed to a product? Sexist distortions become alibis for failure to care. I'm right behind Fiona and asking with her, Why aren't corporate owners and operators of factory farms subjected to the speculation of writers or media commentators about their state of mind, their peculiar biographies, their personal traumas that led them to bring such a horrible site of animal suffering like the factory farm into being?

How do we intervene if we don't understand "damned and dammed desire"? pattrice jones and Cheryl Wylie teach us how to attend to desire and identity, for instance as they function in 4-H programs and backyard hen keeping, so that we can intervene appropriately.

Lori reminds us that this affective turn is an intervening in and against political and literal invocations of domination. Out of disaffection comes an affective and effective turn.

Displace

Where do we place ourselves? How often do we feel displaced? Katie placing herself at auctions, for instance, out of place and in place simultaneously. Yvette Watt chose to create a performance art piece as a specific place to displace the functioning of hunters with guns, to displace the ducks so that they could be safe.

She needed a literal platform to do so—creating space where there wasn't to provide a platform for undermining hunting. In this collaborative, participatory, event-based placement she foregrounded social issues and political activism. She brought people to a space and created protest in this space.

There is a ground to our work.

Conversely, what are places characterized by lack of safety?

Displacement often occurs at the outer edge of society as Liz Bowen explains. What happens there when there are intersections of different kinds of boundaries/loss of boundaries for women and animals?

Discourse

Discourses frame and limit. Discourses don't just interpret, but shape how relationships can and should occur. Liz reminds us that "discourses used to denigrate animal lovers are screens for disdain toward other identity categories." The discourse of personhood that denies animals their claim to personhood is ableist. "Humane" farming narratives create a discourse in which animal victims become seen as partners who collude in their own deaths. Nekeisha Alexis reminds us how conveniently this discourse dispossesses the killer of power.

Alice Crary tackles Peter Singer's discourse on cognitively disabled human beings, while reminding us that even though classic animal comparisons intend to degrade specific human beings, animal comparisons do not necessarily degrade human beings. Is it not time to say that the harm Singer continues to cause by maintaining the importance for his argument of the diminished moral standing of the cognitively impaired shows how he puts a philosophical approach, utilitarianism, before the materiality of any of us? As Alice indicates, Singer is sanguine that he is working with accurate empirical descriptions of the lives of the cognitively disabled. But he should not be. Here is analytic philosophical discourse at its most self-satisfied in its failure to be self-critical.

Silence is both a tool of oppression and a mode of poetic expression, as Hayley Singer suggests. Let us ask of any discourse, who is silent? Who is speaking? Liberation from silence is a storytelling process. But silence can also be a "politically potent form of poetic expression."

Let us pursue ways of co-constructing discourses to balance silence and speaking, to avoid being the "voices for the voiceless" and privileging voice.

Distort

Distortion is everywhere: Ableist metaphors to interpret the existence of speciesism, like equating moral attention with sight and moral inattention with "blindness." Or that humans who don't respond to animal concerns are deaf. Using stereotypes about autism to explain speciesism and then pathologizing the stereotype of autistic people being closer to animals. Saying speciesists are schizophrenic (and then defending it and persisting in it). I was startled to realize how widespread this practice is in anti-speciesist writings. Guy Scotton calls it the "diagnostic impulse to frame speciesism." We also have at work neurotypical people distorting and misunderstanding neurodiversity. One of the results, as this anthology makes clear, is that this keeps us from understanding how definitions of the human are distorted. We are not just responding to "what animals are like"; we are actually in part *creating* "what animals are like" (Scotton quoting Palmer). As Guy puts it, "This array of metaphors constitutes a diagnostic tendency in animal liberation theory and advocacy—a tendency to construe struggles over medical frames of references as a battle to be won rather than a paradigm to be challenged in its own right."

Our task is clear: to call out every new or continuing reference that uses autism, schizophrenia, blindness to explain speciesism and embrace an interspecies concept of neurodiversity.

Distress and disavowal

I am not okay. As Katie says, "I will probably, in some ways, never be OK again." How do we live after being no longer okay? She continued to write her dissertation. She discovered public tears. We must be those who embrace those who are not okay. We must be those who understand we are not okay. It is okay not be okay. The distortion would be to make peace with a world that is not okay.

Former vegans disavow veganism by pathologizing it. Mislabeling veganism with a non-disease concocted out of free-associating it with "anorexia." What's going on there? Thankfully the Stanescu brothers bring us up to speed: "one can refrain from eating animals only as long as no one knows, as long as one repeatedly disavowals any identity or judgment." Individual grief is part of a larger context—collective political action. It is our avowal that helps us survive the grief that accompanies our knowledge. Our avowal about bodies.

Distress and disorientation

Reading this manuscript during late January and early February 2018 helped me recognize the animaladies situation for survivors of sexual harassment and exploitation in the animal rights movement in the United States. Stories of several women who had been sexually harassed over many years began surfacing publicly in a number of papers. These revelations often left those who had already suffered exposed again. I listened to a number of their stories and they felt vulnerable, they didn't know who to trust. Several survivors expressed deep panic about retaliation. They felt exposed.

Then came the disorienting experience—we needed to forgive the men. All the good they had done for animals! The exploiters' apologists wanted to skip the accountability stage. *Their* discomfort with holding abusers accountable meant it was time to "move on." They never discussed the safety of the survivors. #ARMeToo and #TimesUpAR were being pathologized, not the decision of abusive men to endanger women and their organizations' work for animals by exploiting their position of power. *We*—survivors and their advocates–were the problem; according to one prominent funder *we* were engaged in cyber bullying. *We* caused people to be afraid to speak up. What of the women who had been silenced for years by slander suits by abusers against their victims; human resources departments that did not act on their reports of sexual exploitation and hostile workplaces; nondisclosure agreements and threats that they would find no place to work in the animal movement? Women had literally disappeared from the animal rights movement because of what had happened to them. A process of accountability for the perpetrators' actions had only begun (if it had begun at all), but still the survivors were told they needed to "move on." And what of the fact that perpetrators took advantage of women who cared about animals? Their bodies—and their caring—had been exploited.

Distraught

You want to cry for all that has been suffered and you want to scream because it was so much worse than all that you had heard over the years and you want to do something—you want to do *everything*. You don't want despair to take over and you want to congratulate the brave women and you mourn the women who left and you think, "how might the movement have looked if for a quarter of century women weren't told to be quiet, or lost their jobs, or left with disgust, and so weren't part of shaping the movement?"

And how do you explain what feels like a certain instability in your being and the inability to focus and the grasping for something solid and it keeps disappearing? And you realize that you are in the midst of an *animaladies* experience. *This* book is what explains it.

Distribution

I'd love to see this book used in study groups of activists—feminist, disability, and especially animal activists. We have to find ways to take care as we encounter this exciting and challenging material. But we need to make sure it finds its audience. If we care about animals in this patriarchal ableist world, is it not inevitable that each one of us will experience an *animaladies* moment—if not many?

Dis/closure

We are overthrowing so much. We aren't seeking a cure, not according to a patriarchal, speciesist, ableist world. And we aren't going to have closure. Nothing will easily be tied up together; no narrative will find its "finis." But together we can work for transformations.

References

Nochlin, L. (2002), "Joan Mitchell: A Rage to Paint," in J. Livingston, L. Nochlin, and Y. Lee (eds.), *The Paintings of Joan Mitchell*, Berkeley: University of California Press.

Smith, B. G. (2000), *The Gender of History: Men, Women, and Historical Practice*, Cambridge: Harvard University Press.

Thomas, K. (1971), *Religion and the Decline of Magic: Studies in Popular Beliefs in Sixteenth and Seventeenth Century England*, London: Weidenfeld and Nicolson.

INDEX

"old maids" 161–2
Olsen, Tillie 71
oppressions, connected 90–1
orthorexia 5
orthorexia nervosa
    criteria for 138
    diagnosis of 137–40
    indicators of 138
    in theory 138
O'Sullivan, Siobhan 182

Pachirat, Timothy 182
Palas Por Pistolas 211
"paleo diet" 139
Palmer, Clare 110
panic disorder 159
Pateman, Carole 108
pathologization 2–3
patriarchal psychology 105
Patricia 12
Patronek, Gary 178
Perry, Hart 53–4
PerthNow 137
PETA 210
phenomenological empathy 29
physical disabilities 6
pigs 50, 124, 183, 187, 223
    in advertisements and children's
      books 25
    production projects 191
plantation romances 48
    analysis of 56
    conscious omnivore storytelling and
      55–6
    emphasizing affect 57–8
    reframing coercion and violence 60
    representing victims as partners
      59–60
    selective comparing and contrasting
      58–9
    shared tactics of evasion in 56–7
Planter's Northern Bride, The (Hentz) 48,
    57, 60
Plumwood, Val 66
Pollan, Michael 141, 143–6
Polyface Farms 50
Pornography of Meat, The (Adams) 66
"post-autistic economics" movement 106

post-traumatic stress disorder 159
    in animal 25
*Precarious Life: The Powers of Mourning
    and Violence* 74
Prince-Hughes, Dawn 91, 95
Priscilla Queen of the Desert (Elliott) 203
Probyn-Rapsey, Fiona 35, 175–84, 237–8
profound loneliness 79, 81
"profound mental retardation" 118
Project Row Houses 206
psychological disorders 6
Psychology Today 13
"Psychosurgery" (Freeman) 12
public grief 74
Puppies and Babies 219
"Puritanism" 141
pussyhat project 5

Queensland Times 137
queering animal liberation 198–200

Racial Contract, The (Mills) 108
racialized conceptions 3, 4
Raising Chickens for Meat (Roland) 52
Ranciere, Jacques 213
Rattling the Cage (Wise) 103
Regan, Tom 4
reproductive tyranny 50
Reyes, Pedro 211
Reynolds, Diamond 15, 16
Rich, S. 161
Rock, M. J. 166
Rogers, Melvin 16
Roland, Gwen 52
Rowling, J. K. 176
Rudy, Kathy 142

Salatin, Joel 50
Salomon, D. 90, 91, 93, 110
Scarry, Elaine 228
schizophrenia 71, 106
    social form of 103
Schneemann, Carolee 219
Scotton, Guy 6, 101–13, 239
sculptural body fragmentation, by
    mowson 25
    babyforms 29
    boobscape 36–41